tredition

tredition was established in 2006 by Sandra Latusseck and Soenke Schulz. Based in Hamburg, Germany, tredition offers publishing solutions to authors and publishing houses, combined with worldwide distribution of printed and digital book content. tredition is uniquely positioned to enable authors and publishing houses to create books on their own terms and without conventional manufacturing risks.

For more information please visit: www.tredition.com

TREDITION CLASSICS

This book is part of the TREDITION CLASSICS series. The creators of this series are united by passion for literature and driven by the intention of making all public domain books available in printed format again - worldwide. Most TREDITION CLASSICS titles have been out of print and off the bookstore shelves for decades. At tredition we believe that a great book never goes out of style and that its value is eternal. Several mostly non-profit literature projects provide content to tredition. To support their good work, tredition donates a portion of the proceeds from each sold copy. As a reader of a TREDITION CLASSICS book, you support our mission to save many of the amazing works of world literature from oblivion. See all available books at www.tredition.com.

Project Gutenberg

The content for this book has been graciously provided by Project Gutenberg. Project Gutenberg is a non-profit organization founded by Michael Hart in 1971 at the University of Illinois. The mission of Project Gutenberg is simple: To encourage the creation and distribution of eBooks. Project Gutenberg is the first and largest collection of public domain eBooks.

The Rain Cloud or, An Account of the Nature, Properties, Dangers and Uses of Rain in Various Parts of the World

Charles Tomlinson

Imprint

This book is part of TREDITION CLASSICS

Author: Charles Tomlinson
Cover design: Buchgut, Berlin – Germany

Publisher: tradition GmbH, Hamburg - Germany
ISBN: 978-3-8472-1651-3

www.tredition.com
www.tredition.de

Copyright:
The content of this book is sourced from the public domain.

The intention of the TREDITION CLASSICS series is to make world literature in the public domain available in printed format. Literary enthusiasts and organizations, such as Project Gutenberg, worldwide have scanned and digitally edited the original texts. tredition has subsequently formatted and redesigned the content into a modern reading layout. Therefore, we cannot guarantee the exact reproduction of the original format of a particular historic edition. Please also note that no modifications have been made to the spelling, therefore it may differ from the orthography used today.

the
RAIN CLOUD:

or,
An Account
of
THE NATURE, PROPERTIES, DANGERS,
AND USES OF RAIN,

in various parts of the world.

published under the direction of
the committee of general literature and education
appointed by the society for promoting
christian knowledge.
LONDON:
Printed for the
society for promoting christian knowledge.
sold at the depository,
great queen street, lincoln's inn fields,
and 4, royal exchange.

1846.

p. 4 london:
r. clay, printer, bread street hill.

CHAPTER I:

rains peculiar to each season—spring showers—midsummer rains—rains of autumn and winter—means of supplying the earth with rain—rain-clouds—deceptive appearances of clouds—their light and shade—effects of clouds in mountainous countries—ascent of monte pientio—ascent to the peak of teneriffe—grand effects of clouds in the pyrenees—voyage in a balloon through the clouds.

Every season has its own peculiar rains. What can be more refreshing or invigorating than the showers of spring? When the snows of February have disappeared, and the blustering winds of March have performed their office of drying up the excess of moisture, and preparing the earth for fruitfulness, and when the young buds and blossoms of April are peeping forth beneath the influence of the sun, and the trees and hedges are attired in their new robes of tender green, how soon would all this beauty languish but for the showers of spring! Several dry days, perhaps, have passed, and the wreaths of dust which are raised by the wind show that the earth wants moisture; but before a drop falls there is a general lull throughout all nature; not a leaf is heard to rustle; the birds are mute and the cattle stand in expectation of the refreshing fall. At last the pools and rivulets are "dimpled" by a few soft drops, the forerunners of the general shower. And this shower, unlike the heavier rains of summer, comes stealing on so gently, that the tinkling sound of its fall is heard among the branches of the bursting trees long before it is felt by those who walk beneath their slight shelter. Rapidly does the landscape brighten under the influence of the welcome shower; and as it becomes more rich and extensive, all nature seems to rise up and rejoice. The birds chirp merrily among the foliage; the flowers raise their drooping heads, and the thirsty ground drinks in with eager haste the mellowing rains. All day long, perhaps, does the rain continue to fall, until the earth is fully moistened and "enriched with vegetable life." At length, towards evening, the sun peeps out from among the broken clouds, and lights up, by his sudden radiance, the lovely scene. Myriads of raindrops sparkle like gems beneath his beams; a soft mist that seems to mingle earth and sky gradually rolls away, and "moist, and bright, and green, the landscape laughs around." Now pours forth the evening concert from the woods, while warbling brooks, and lowing herds, appear to answer to the sound. Such are some of the delightful effects of spring-showers.

In summer, when the heat has been very great, the rain is often ushered in by a thunder-storm, and falls in torrents, which at an earlier season would do harm to the young buds and blossoms of spring; but now the vegetation is strong enough to resist the floods so necessary to maintain moisture in the parched earth. But when the summer has been moderately warm some gentle rains generally fall about midsummer, which, from the frequency of their occurrence about this time, have obtained the name of "Midsummer rains." These rains are popularly associated with St. Swithin's Day, as will be noticed in another chapter; but when they fall early, mildly, and in moderate quantity, they operate to a certain extent as a p. 14second spring. "Many of the birds come into song and have second broods; and it is probable that there is a fresh production of caterpillars for their food, or, at all events, a larger production of the late ones than when the rains are more violent and protracted. Many of the herbaceous plants also bloom anew, and the autumn is long and pleasant, and has very many of the charms of a summer, though without any very powerful operation on the productions of nature, further than a very excellent preparation for the coming year, whether in buds, in roots, or in the labours of man. Such a season is also one of plenty, or at all events of excellent quality in all the productions of the soil. The wild animals partake in the general abundance, as that food which is left for them in the fields, after man has gathered in his share, is both more abundant and more nourishing. When there is much moisture from the protracted time and great quantity of the rains, many of those seeds germinate, while in mild seasons they are left as food for the wild animals, chiefly the field-mice and the birds, which again form part of the food of the predatory ones."

There is something melancholy and depressing p. 15in the rains of autumn and winter, for they bear away the last traces of summer by stripping the trees of the many-coloured leaves, which in mild seasons will continue to adorn the landscape even late in November. The rains of this month, and their effects, have been skilfully sketched by an accurate observer of nature. He says: —

> "Now cold rains come deluging down, till the drenched ground, the dripping trees, the pouring eaves, and the torn, ragged-skirted clouds, seemingly dragged downward slantwise by the threads of dusky rain that descend from them, are all mingled together in one blind confusion; while the few cattle that are left in open pastures, forgetful of their till now interminable business of feeding, turn their backs upon the besieging storm, and, hanging down their heads till their noses almost touch the ground, stand out in the middle of the fields motionless, like dead images.
>
> "Now, too, a single rain-storm, like the above, breaks up all the paths and ways at once, and makes home no longer 'home' to those who are not obliged to leave it; while it becomes doubly endeared to those who are. What sight, for instance, is so pleasant to the wearied woodman, who p. 16has been out all day long in the drenching rains of this month, as his own distant cottage window seen through the thickening dusk, lighted up by the blazing fagot that is to greet his sure return at the accustomed minute?"

While we watch the effects of the various rains, and their beneficial influence on the earth, there is also much to excite our gratitude and admiration; for among the many beautiful contrivances in creation, none is more remarkable than the means by which the earth is watered and refreshed by rain. The oceans, seas, lakes, and other waters of the earth supply the air with moisture, which, rendered elastic and invisible by the heat of the sun and of the earth, rises to various heights in the atmosphere, where it forms clouds in all their wonderful beauty and variety. These are borne by the winds to places far inland, to which water in sufficient quantity could not come by any other means, and where moisture is most required; and here the water is poured down, not in cataracts and waterspouts, but in the form of drops of various sizes. If the rain-clouds threw down, at once and suddenly, all the water contained in them, not only would vegetation be destroyed by the force of the fall, but we should be constantly liable to floods and other inconveniences. Clouds also serve to screen the earth from the fierce heat of the sun by day; and, by night, they serve to maintain the heat which would otherwise escape by radiation, and produce great cold even in summer. Clouds thus have great influence in regulating the extremes of heat and cold, and in forming what is called the "climate" of a country. Clouds also supply the hidden stores of fountains and the fresh water of rivers; and, as a pious old divine well remarks, "So abundant is this great blessing, which the most indulgent Creator hath afforded us by means of this distribution of the waters I am speaking of, that there is more than a scanty, bare provision, a mere sufficiency; even a plenty, a surplusage of this useful creature of God, the fresh waters afforded to the world; and they so well ordered, as not to drown the nations of the earth, nor to stagnate, stink, and poison, or annoy them; but to be gently carried through convenient channels back again to their grand fountain the sea; and many of them through such large tracts of land and to such prodigious distances, that it is a great wonder the fountains should be high enough, or the seas low enough, ever to afford so long a conveyance." [18]

If rain is not at all seasons pleasant and delightful, neither are rain-clouds among the most beautiful which diversify the landscape of the sky; for it has been well remarked, that "all the fine-weather clouds are beautiful, and those connected with rain and wind mostly the reverse." What, indeed, can be more striking than the akrial landscapes of fine weather, in which, by an easy fancy, we can trace trees and towers, magnificent ruins and glaciers, natural bridges and palaces, all dashed with torrents of light or frowning in shadow, glowing like burnished silver, glittering in a golden light, or melting into the most enchanting hues? But with all this beauty the eye is seldom capable of judging correctly of the proper size and forms and motions of clouds. The same cloud which to one observer may be glowing with light, to another may be enveloped in shadow. That which appears to be its summit may be only a portion of its outer edge, while that which seems to be its lower bed may really be a portion of its further border. A spectator, on the summit of a p. 19tall cliff, may observe what he takes to be a single cloud; while a second spectator, on lower ground, will perceive that there are two clouds. The motions of clouds are so deceptive, that they often seem to be moving in a curve over the great concave of heaven, while they are in fact advancing in nearly a right line. Suppose, for example that a cloud is moving from the distant horizon p. 20towards the place where we stand, in a uniform horizontal line without changing either in size or form. Such a cloud, when first seen, will appear to be in contact with the distant horizon, and consequently much nearer to us than it really is. As it advances towards us, it will seem to rise into the sky, and to become gradually larger till it is almost directly overhead. Continuing its progress, it will then seem again to descend and to lessen in size as gradually as it had before increased; till at length it disappears in the distant horizon at a point exactly opposite to that at which it was first seen. Thus the same cloud, without varying its motion in the least from a straight line, and remaining throughout of the same size and form, would seem to be continually varying in magnitude; and the line of its motion, instead of being straight, would appear to be curved. This is one of the most simple cases that can be supposed: but the clouds as they exist in nature do not remain of the same magnitude, but are constantly changing in form, in size, in direction, and in velocity; so that it is quite impossible to form an accurate idea of their shape and size, or to explain their motions. Clouds, at different elevations, p. 21may often be seen to move in different directions under the influence of different currents of wind.

The distribution of light and shade in clouds is most striking. The watery particles of which they are composed, yielding constantly to changes in temperature and moisture, are always changing; so that a most beautiful cloud may alter in figure and appearance in an instant of time; the light parts may suddenly become dark, and those that were shaded may all at once glow in the rays of the sun. Again, the appearance of a cloud, with respect to the sun, may entirely alter its character. The same cloud, to one observer, may appear entirely in shade, to another tipped with silver; to a third it may present brilliant points and various degrees of shade, or one of its edges only may appear illuminated; sometimes the middle parts may appear in shadow, while the margin may be partially luminous, rendering the middle parts all the more obscure by the contrast.

A wonderful variety may also be produced by the shadow of one cloud falling upon another. The accompanying sketch furnishes an example of this. Sometimes the whole of a cloud projects a shadow through the air upon some other far p. 22distant cloud, and this again upon another, until at length it reaches the ground. The shadows of moving clouds may often be traced upon the ground, and they contribute greatly to modify the appearance of the landscape. A large number of small flickering clouds produce broken lights and shades which have an unpleasant jarring effect; but when the clouds are massive, or properly distributed, p. 23the shadows often produce a high degree of repose.

Clouds are often seen to advantage in mountainous countries. Here the aspect of the heavens may be entirely different at different elevations. A single cloud in the valley may conceal the whole of the upper sky from an observer; but as he ascends he may gradually get above this and other layers or bands of cloud, and see a beautifully variegated sky above him, while the clouds which conceal the valley may be rolling at his feet. Evelyn, in his Memoirs, notices a scene of this kind. He says,—"Next morning we rode by Monte Pientio, or, as vulgarly called, Monte Mantumiato, which is of an excessive height, ever and anon peeping above airy clouds with its snowy head, till we had climbed to the inn at Radicofany, built by Ferdinand the greate Duke for the necessary refreshment of travellers in so inhospitable a place. As we ascended we entered a very thick, solid, and dark body of cloudes, which looked like rocks at a little distance, which lasted neare a mile in going up; they were dry, misty vapours, hanging undissolved for a vast thicknesse, and obscuring both sun and earth, so that we p. 24seemed to be in the sea rather than in the cloudes, till, having pierced through it, we came into a most serene heaven, as if we had been above all human conversation, the mountain appearing more like a great island than joyn'd to any other hills, for we could perceive nothing but a sea of very thick cloudes rowling under our feete like huge waves, every now and then suffering the top of some other mountain to peepe through, which we could discover many miles off: and betweene some breaches of the cloudes we could see landskips and villages of the subjacent country. This was one of the most pleasant, newe, and altogether surprising objects that I had ever beheld."

In the following interesting account of the ascent of the Peak of Teneriffe by Captain Basil Hall, it will be seen that heavy rain clouds may skirt the mountain, while its summit is in a pure and dry air.

> "On the 24th of August," he says, "we left Oratava to ascend the Peak. The day was the worst possible for our purpose, as it rained hard; and was so very foggy that we could not see the Peak, or indeed any object beyond one hundred yards distant.
>
> p. 25"After riding slowly up a rugged path for four hours, it became extremely cold, and, as the rain never ceased for an instant, we were by this time drenched to the skin, and looked with no very agreeable feelings to the prospect of passing the night in wet clothes. At length the night began to close in, and the guides talked of the improbability of reaching the English station before night. It was still raining hard; but we dismounted, and took our dinner as cheerfully as possible, and hoping for clearer weather the next day. On remounting, we soon discovered that the road was no longer so

steep as it had been heretofore, and the surface was comparatively smooth: we discovered, in short, that we had reached a sort of table-land, along which we rode with ease. Presently we thought the fog less dense, and the drops of rain not so large, and the air less chilling. In about half an hour we got an occasional glimpse of the blue sky; and as we ascended, (for our road, though comparatively level, was still upon the rise,) these symptoms became more manifest. The moon was at the full, and her light now became distinct, and we could see the stars in the zenith. By this time we had reached the Llano p. 26de los Remenos, or Retamos Plain, which is many thousand feet above the sea; and we could distinctly see that during the day we had merely been in a cloud, above which having now ascended, the upper surface lay beneath us like a country covered with snow. It was evident, on looking round, that no rain had fallen on the pumice gravel over which we were travelling. The mules were much fatigued, and we got off to walk. In a few minutes our stockings and shoes were completely dried, and in less than half an hour all our clothes were thoroughly dried. The air was sharp and clear, like that of a cold frosty morning in England; and though the extreme dryness, and the consequent rapid evaporation, caused considerable cold, we were enabled by quick exercise to keep ourselves comfortable. I had various instruments with me, but no regular hygrometer: accident, however, furnished me with one sufficiently indicative of the dry state of the air. My gloves, which I kept on while mounted, were completely soaked with the rain; and I took them off during this walk, and, without considering what was likely to happen, rolled them up, and carried them in my hand. When, at the end of an hour, or somewhat p. 27less, we came to remount our mules, I found the gloves as thoroughly dried and shrivelled up as if they had been placed in an oven. During all the time we were at the Peak itself, on the 26th, the sky was clear, the air quite dry, and we could distinguish, several thousand feet below us, the upper and level surface of the stratum of clouds through which we had passed the day before, and into which we again entered on going down, and found precisely in the same state as when we started."

It is not uncommon to observe an effect quite contrary to the one given in the last two examples, the high summits of mountains being frequently concealed by heavy clouds of mist, while at a very short distance below them the air is clear and pure. In ascending to the Port of Venasque, one of the mountain passes of the Pyrenees, Mr. Murray found the mists so dense that he despaired of getting above them, or of their clearing away. But fortunately the wind freshened, and the mist, broken by it, "came sweeping," he says, "over our heads, sometimes enveloping us in darkness, sometimes exposing the blue sky, and a part of the mountains. Section after section of the bald and towering masses which rose above the path p. 28were displayed to us, one after another, as if the whole had been a sight too great for us to look upon. Sometimes the clouds opened, and the snows, sparkling in the sun-beams, were before us; at others, an enormous peak of the mountain would shoot its dark head through the mist, and, without visible support, seem as if it were about to fall upon us. Again, when we imagined ourselves hemmed in on all sides by the mountains, and within a few feet of their rugged sides, a passing breeze would disclose the dark waters of the lakes hundreds of feet beneath us.

"Thus the effect of light and darkness, of sunshine and of mist, working upon materials of such grandeur as those near the Port of Venasque, was a sight well worthy of admiration, and one which is rarely to be seen. * * * * Excepting the intervals of light which the gusts of wind, by dispersing the mists, had bestowed upon us, we had hitherto, comparatively speaking, been shrouded in darkness, particularly for the ten minutes preceding our arrival at the Port: my astonishment may therefore be imagined when, the instant that I stepped beyond the limits of the Port, I stood in the purest atmosphere—not a particle of mist, p. 29not even a cloud, was perceptible. The phenomenon was curious, and its interest greatly heightened from the situation in which it took place. The mist rolling up the valley through which we had passed, was, the moment that it could be said to reach the Spanish frontier,—the moment it encircled the edges of the high ridges which separated the countries, thrown back, as it were, indignantly, by a counter current from the Spanish side. The conflicting currents of air, seemingly of equal strength, and unable to overcome each other, carried the mist perpendicularly from the summits of the ridge, and filling up the crevices and fissures in its uneven surface, formed a wall many thousand feet above it, of dark and (from the appearance of solidity which its massive and perpendicular character bestowed upon it) apparently impenetrable matter."

Undoubtedly the various phenomena of clouds may be seen to great advantage in mountain regions; and there is only one other method of seeing them to greater perfection, and that is from the car of a balloon. The following description of an akrial voyage, by Mr. M. Mason, in October 1836, will convey a better idea of the magnificence p. 30of a cloudy sky than any terrestrial prospect could do. He says,—

> "Scarcely had we quitted the earth before the clouds, which had previously overhung us, began to envelop us on all sides, and gradually to exclude the fading prospect from our sight. It is scarcely possible to convey an adequate idea of the effect produced by this apparently trivial occurrence. Unconscious of our own motion from any direct impression upon our own feelings, the whole world appeared to be in the act of receding from us in the dim vista of infinite space; while the vapoury curtain seemed to congregate on all sides and cover the retreating masses from our view. The trees and buildings, the spectators and their crowded equipages, and finally, the earth itself, at first distinctly seen, gradually became obscured by the thickening mist, and growing whiter in their forms, and fainter in their outlines, soon faded

away 'like the baseless fabric of a vision,' leaving us, to all appearance, stationary in the cloud that still continued to involve us in its watery folds. To heighten the interest and maintain the illusion of the scene, the shouts and voices of the multitude whom we had left behind us, cheering the p. 31ascent, continued to assail us, (long after the interposing clouds had effectually concealed them from our eyes,) in accents which every moment became fainter and fainter till they were finally lost in the increasing distance.

"Through this dense body of vapour, which may be said to have commenced at an altitude of about 1000 feet, we were borne upwards to perhaps an equal distance, when the increasing light warned us of our approach to its superior limits, and shortly after, the sun and we rising together, a scene of splendour and magnificence suddenly burst upon our view, which it would be vain to expect to render intelligible by any mode of description within our power. Pursuing the illusion, which the previous events had been so strongly calculated to create, the impression upon our senses was that of entering upon a new world to which we had hitherto been strangers, and in which not a vestige could be perceived to remind us of that we had left, except the last faint echo of the voices which still dimly reached us, as if out of some interminable abyss into which they were fast retreating.

"Above us not a single cloud appeared to disfigure p. 32the clear blue sky, in which the sun on one side, and the moon in her first quarter on the other, reigned in undisturbed tranquillity. Beneath us, in every direction, as far as the eye could trace, and doubtless much further, the whole plane of vision was one extended ocean of foam, broken into a thousand fantastic forms; here swelling into mountains, there sinking into lengthened fosses, or exhibiting the appearance of vast whirlpools; with such a perfect mimicry of the real forms of nature, that, were it not for a previous acquaintance with the general character of the country below us, we should frequently have been tempted to assert, without hesitation, the existence of mountainous islands penetrating through the clouds, and stretching in protracted ranges along the distant verge of our horizon.

"In the centre of this hemisphere, and at an elevation of about 3000 feet above the surface of the clouds, we continued to float in solitary magnificence; attended only at first by our counterpart—a vast image of the balloon itself with all its paraphernalia distinctly thrown by the sun upon the opposite masses of vapour, until we had risen so high that even that, outreaching the material p. 33basis of its support, at length deserted us; nor did we again perceive it until, preparatory to our final descent, we had sunk to a proper elevation to admit of its re-appearance.

"Not the least striking feature of our, and similar situations, is the total absence of all perceptible motion, as well as of the sound which, in ordinary cases, is ever found to accompany it. Silence and tranquillity appear to hold equal and undisputed sway throughout these airy regions. No matter what may be the convulsions to which the atmosphere is subjected, nor how violent its effects in sound and motion upon the agitated surface of the earth, not the slightest sensation of either can be detected by the individual who is floating in its currents. The most violent storm, the most outrageous hurricane, pass equally unheeded and unfelt; and it is only by observing the retreating forms of the stable world beneath, that any certain indication can be obtained as to the amount or violence of the motion to which the individual is actually subjected. This, however; was a resource of which we were unable to avail ourselves, totally excluded as we were from all view of the earth, or any fixed point connected with it.

p. 34"Once, and only once, for a few moments preparatory to our final descent, did we obtain a transitory glimpse of the world beneath us. Upon approaching the upper surface of the vapoury strata, which we have described as extending in every direction around, a partial opening in the clouds discovered to us for an instant a portion of the earth, appearing as if dimly seen through a vast pictorial tube, rapidly receding behind us, variegated with furrows, and intersected with roads running in all directions; the whole reduced to a scale of almost graphic minuteness, and from the fleecy vapour that still partially obscured it, impressing the beholder with the idea of a vision of enchantment, which some kindly genius had, for an instant, consented to disclose. Scarcely had we time to snatch a hasty glance, ere we had passed over the spot, and the clouds uniting gradually concealed it from our view.

"After continuing for a short space further, in the vain hope of being again favoured with a similar prospect, the approach of night made it desirable that we should prepare for our return to earth, which we proceeded to accomplish accordingly."

p. 36

CHAPTER II.

effects of rain in mountainous districts—the district of moray—the great floods of 1829—commencement of the rain—the swollen rivers—disastrous effects of the flood—means adopted for the rescue of cottagers—kerr and his brave deliverers—rescue of funns and his family—floods of the rhone in 1840—overflowing of the mississippi.

It is well known that some years are wetter than others; but to persons living in tolerably flat countries an unusually wet season causes no great inconvenience. It interferes, it is true, with outdoor employments, but people seldom apprehend any danger from the long continuance of rain. It is not so, however, in hilly or mountainous regions; an unusual fall of rain swells the rivers to such an extent, that they often overflow their banks, and occasion much damage to the surrounding districts; or, where the river's banks are defended on both sides by perpendicular rocks, the waters sometimes rise so fast as to attain a height of forty or fifty feet above their natural level, and from this height they pour with destructive violence over the face of the country. Such was the case in the great floods of Moray, which happened in the year 1829, of which the following is a brief abstract, derived chiefly from Sir Thomas Dick Lauder's interesting volume on this subject, published soon after the calamity for the benefit of the sufferers.

The province of Moray, or Murray, is a large district in the north-east of Scotland, bounded by the Moray Frith on the north-east and north. The eastern half of the province is lower than the western; in which the mountains render the whole country characteristically highland. On the north is a long belt of lowlands, about 240 square miles in extent: this is greatly diversified with ridgy swells and low hilly ranges, lying parallel to the frith, and intersected by the rivers Ness, Nairn, Findhorn, Lossie, and Spey running across it to the sea. The grounds behind the lowlands appear, as seen from the coast, to be only a narrow ridge of bold alpine heights, rising like a rampart to guard the orchards, and woods, and fields: but these really form long and broad mountain masses, receding, in all the wildness and intricacy of highland arrangement, to a distant summit line.

Some of the broad clifts and long narrow vales of these mountains form beautiful and romantic pictures; while many of their declivities are practicable to the plough or other instruments of cultivation; so that the bottoms and the reclaimed or reclaimable sides of the valleys are estimated to comprehend about one-third of the entire area. The lowlands of Moray have long been celebrated for mildness and luxuriousness of climate, and also for a certain dryness of atmosphere, which seems to have some intimate connexion with the mournful calamity about to be described. The high broad range of mountains on the south-west shelter the lowlands from the prevailing winds of the country, and exhaust many light vapours and thinly-charged clouds, which might otherwise produce gentle rains; but, for just the same reason, they powerfully attract whatever long broad streams of heavy clouds are sailing through the sky, and, among the gullies and the upland glens, amass their discharged contents with amazing rapidity, and in singular largeness of volume. The rivers of the country are, in consequence, peculiarly liable to become flooded. One general and tremendous outbreak, p. 40in 1829, "afforded an awful exhibition of the peculiarities of the climate, and will long be remembered, in connexion with the boasted luxuriousness of Moray, as an illustration of how chastisement and comfort are blended in a state of things which is benignly adjusted for the moral discipline of man, and the correction of moral evil."

The heat in the province of Moray during the summer of 1829 was unusually great. In May the drought was so excessive, as to kill many of the recently planted shrubs and trees. As the season advanced, the variations in the barometer became so remarkable, that observers began to lose all confidence in this instrument.

The deluge of rain, which produced the flood of the 3d and 4th of August, fell chiefly on the Monadhlradh mountains, rising between the south-east part of Lochness and Kingussie, in Badenoch, and on that part of the Grampian range forming the somewhat independent groups of the Cairngorums. The westerly winds, which prevailed for some time previously, seem to have produced a gradual accumulation of vapour to the north of our island, and the column, being suddenly impelled by a strong north-easterly blast, was driven p. 41towards the south-west, its right flank almost sweeping the Caithness and Sutherland coasts, until rushing up and across the

Moray Frith it was attracted by the lofty mountains just mentioned, and discharged in fearful torrents. There fell at a great distance from the mountains, within twenty-four hours, about one-sixth of the annual allowance of rain; on the mountains themselves the deluge that descended, must have been so enormous as to occasion surprise that a flood, even yet more tremendous in its magnitude and consequences, did not result from it.

The mouth of the Findhorn is described as the most important scene of action. The banks of this river are well defended by rocks on either side, and its whole course is distinguished by the most romantic scenery. At the part where it is crossed by the old military bridge of Dulsie, the scenery is of the wildest character. The flood was most tremendous at this bridge, for the water was so confined that it filled the smaller arch altogether, and rose in the great arch to within three feet of the key-stone, that is to say, forty feet above the usual level. This fine old bridge sustained but little damage, while many of the modern buildings p. 42were entirely swept away. At another part of the river, it is stated, as a curious illustration of the height to which the stream had risen, that a gardener waded into the water as it had begun to ebb on the haugh, and with his umbrella drove ashore and captured a fine salmon, at an elevation of fifty feet above the ordinary level of the Findhorn.

At Randolph's bridge the opening expands as the rocks rise upwards, till the width is about seventy or eighty feet; yet, from the sudden turn of the river, as it enters this passage, the stream was so checked in its progress that the flood actually rose over the very top of the rocks, forty-six feet above the usual height, and inundated the level part that lies over them to the depth of four feet, making a total perpendicular rise at this point of not less than fifty feet.

The effects of the deluge of the 3d and 4th of August, remain on the Dorbach, in a bank one hundred feet high, which rose with slopes and terraces covered with birch and alder wood. The soil being naturally spongy imbibed so much rain, that it became overloaded, and a mass of about an acre in extent, with all its trees on it, gave way at p. 43once, threw itself headlong down, and bounded across the bed of the Dorbach, blocking up the waters, flooded and wide as they were at the time. A farmer, who witnessed this phenomenon, told Sir Thomas Dick Lauder that it fell "wi' a sort o' a dumb sound," while astonished and confounded p. 44he remained gazing at it. The bottom of the valley is here some two hundred yards or more wide, and the flood nearly filled it. The stoppage was not so great, therefore, as altogether to arrest the progress of the stream; but this sudden obstacle created an accumulation of water behind it, which went on increasing for nearly an hour, till, becoming too powerful to be longer resisted, the enormous dam began to yield, and was swept off at once, and hurled onwards like a floating island. While the farmer stood lost in wonder to behold his farm thus sailing off to the ocean by acres at a time, another half acre, or more, was suddenly rent from its native hill, and descended at once, with a whole grove of trees on it, to the river, where it rested on its natural base. The flood immediately assailed this, and carried off the greater part of it piecemeal. At the time when Sir Thomas was writing, part of it remained with the trees growing on it in the upright position, after having travelled through a horizontal distance of sixty or seventy yards, with a perpendicular descent of not less than sixty feet.

At Dunphail, the residence of Mr. Bruce was threatened by the flood, and that gentleman prevailed p. 45on his wife and daughter to quit the house and seek refuge on higher ground. Before quitting the place, their anxiety had been extremely excited for the fate of a favourite old pony, then at pasture in a broad green, and partially-wooded island, of some acres in extent. As the spot had never been flooded in the memory of man, no one thought of removing the pony until the wooden bridges having been washed away rendered it impossible to do so. When the embankment gave way, and the patches of green gradually diminished, Dobbin, now in his 27th year, and in shape something like a 74-gun ship cut down to a frigate, was seen galloping about in great alarm as the wreck of roots and trees floated past him, and as the last spot of grass disappeared he was given up for lost. At this moment he made a desperate effort to cross the stream under the house; the force of the current turned him head over heels, but he rose again with his head up the river; he made boldly up against it, but was again borne down and turned over: every one believed him lost, when rising once more and setting down the waste of water, he crossed both torrents, and landed safely on the opposite bank.

p. 46At night Mr. Bruce says there was something inexpressibly fearful and sublime in the roar of the torrent, which by this time filled the valley, the ceaseless plash of the rain, and the frequent and fitful gusts of the north wind that groaned among the woods. The river had now undermined the bank the house stood on, and this bank had already been carried away to within four paces of the foundation of the kitchen tower, and, as mass after mass fell with a thundering noise, some fine trees, which had stood for more than a century on the terrace above it, disappeared in the stream. The operations of the flood were only dimly discovered by throwing the faint light of lanterns over its waters, and its progress was judged of by marking certain intervals of what remained of the terrace. One by one these fell in, and at about eleven o'clock the river was still rising, and only a space of three yards remained about the house, which was now considered as lost. The furniture was ordered to be removed, and by means of carts and lanterns this was done without any loss. About one o'clock in the morning, the partial subsidence of the flood awakened a slight hope, but in an hour it rose again higher than before. The banks p. 47which supported the house were washed away, and the house itself seemed to be doomed, and the people were therefore sent out of it. But Providence ordered otherwise; about four o'clock the clouds appeared higher, the river began again to subside; by degrees a little sloping beach became visible towards the foot of the precipice; the flood ceased to undermine, and the house was saved.

But the ruin and devastation of the place were frightful to behold. The shrubbery, all along the river side, with its little hill and moss-house, had vanished; two stone and three wooden buildings were carried off; the beautiful fringe of wood on both sides of the river, with the ground it grew on, were washed to the ocean, together with all those sweet and pastoral projections of the fields which gave so peaceful and fertile a character to the valley; whilst the once green island, robbed of its groups of trees and furrowed by a dozen channels, was covered with large stones, gravel, and torn-up roots.

At another part of the same river (the Divie) Sir Thomas describes, from his own observations, the progress of the flood. The noise was a distinct combination of two kinds of sound: one, an uniform continued roar; the other, like rapidly repeated discharges of cannon. The first of these proceeded from the violence of the water; the other, which was heard through it, and as it were muffled by it, came from the numerous stones which the stream was hurling over its uneven bed of rock. Above all this was heard the shrieking of the wind. The leaves were stripped off the trees and whirled into the air, and their thick boughs and stems were bending and cracking beneath the tempest. The rain was descending in sheets, not in drops: and a peculiar lurid, bronze-like hue pervaded the whole face of nature. And now the magnificent trees were overthrown faster and faster, offering no more resistance than reeds before the mower's scythe. Numerous as they were, they were all, individually, well-known friends. Each, as it fell, gave one enormous plash on the surface, then a plunge, the root upwards above water for a moment; again all was submerged—and then up rose the stem disbranched and peeled; after which, they either toiled round in the cauldron, or darted, like arrows, down the stream. "A chill ran through our hearts as we beheld how rapidly the ruin of our favourite and long-cherished spot was going on. But we remembered that the calamity came from the hand of God; and seeing that no human power could avail, we prepared ourselves to watch every circumstance of the spectacle." In the morning the place was seen cleared completely of shrubs, trees, and soil; and the space so lately filled with a wilderness of verdure was now one vast and powerful red-coloured river.

On the left bank of the Findhorn the discharge of water, wreck, and stones that burst over the extensive plain of Forres, spreading devastation abroad on a rich and beautiful country, was truly terrific. On the 3d of August, Dr. Brands, of Forres, having occasion to go to the western side of the river, forded it on horseback, but ere he crossed the second branch of the stream, he saw the flood coming thundering down. His horse was caught by it; he was compelled to swim; and he had not long touched dry land ere the river had risen six feet. By the time he had reached Moy the river had branched out into numerous streams, and soon came rolling on in awful grandeur; the effect being greatly heightened by the contrary direction of the northerly p. 50wind, then blowing a gale. Many of the cottages occupied a low level, and the inhabitants were urged to quit them. Most of them did so; but some, trusting to their apparent distance from the river, refused to move.

About ten o'clock the river had risen and washed away several of the cottages; and on every side were heard reports of suffering cottagers, whose houses were surrounded by water. One of them was Sandy Smith, an active boatman, commonly called *Whins*, (or *Funns*, as it is pronounced,) from his residence on a piece of furzy pasture, at no great distance from the river. From the situation of his dwelling he was given up for lost; but for a long time the far-distant gleam of light that issued from his window showed that he yet lived.

The barns on the higher grounds accommodated many people; and large quantities of brose (broth) were made for the dripping and shivering wretches. Candles were placed in all the windows of the principal house (that of Mr. Suter) that poor Funns might see he was not forgotten. But, alas! his light no longer burns, and in the midst of the tempest and darkness, it was utterly vain to attempt to assist the distressed.

p. 51At daybreak the wide waste of waters was only bounded by the rising grounds on the south and west: whilst, towards the north and east, the watery world swept off, uninterruptedly, into the expanding Frith and the German Ocean. The embankments appeared to have everywhere given way; and the water that covered the fields, lately so beautiful with yellow wheat, green turnips, and other crops, rushed with so great impetuosity in certain directions, as to form numerous currents, setting furiously through the quieter parts of the inundation, and elevated several feet above it. As far as the eye could reach the brownish-yellow moving mass of water was covered with trees and wreck of every description, whirled along with a force that shivered many of them against unseen obstacles. There was a sublimity in the mighty power and deafening roar of waters, heightened by the livid hue of the clouds, the sheeting rain, the howling of the wind, the lowing of the cattle, and the screaming and wailing of the assembled people, that riveted the attention. In the distance could dimly be descried the far-off dwelling of poor Funns, its roof rising like a speck above the flood, that had evidently made a breach in one of its ends.

p. 52A family named Kerr, who had refused to quit their dwelling, were the objects of great anxiety. Their son, Alexander Kerr, had been watching all night, and in the morning was still gazing towards the spot in an agony of mind, and weeping for the apparently inevitable destruction of his parents. His master tried to comfort him; but even whilst he spoke, the whole gable of Kerr's dwelling, which was the uppermost of three houses composing the row, gave way, and fell into the raging current. Dr. Brands, who was looking on intently at the time, with a telescope, observed a hand thrust through the thatch of the central house. It worked busily, as if in despair of life; a head soon appeared; and at last Kerr's whole frame emerged on the roof, and he began to exert himself in drawing out his wife and niece. Clinging to one another, they crawled along the roof towards the northern chimney. The sight was torturing. Kerr, a little a-head of the others, was seen tearing off the thatch, as if trying to force an entrance through the roof, whilst the miserable women clung to the house-top, the blankets which they had used to shelter them almost torn from them by the violence of the p. 53hurricane; and the roof they had left yielding and tottering, fell into the sweeping flood. The thatch resisted all Kerr's efforts; and he was now seen to let himself drop from the eaves on a small speck of ground higher than the rest, close to the foundation of the back wall of the buildings, which was next the spectators. There he finally succeeded in bringing down the women; and there he and they stood, without even room to move.

p. 54Some people went on horseback to try to procure boats. They managed to get on some way by keeping the line of road. The water was so deep that the horses were frequently swimming; but at length the current became so strong that they were compelled to seek the rising grounds. Dr. Brands attempted to reach the bridge of Findhorn, in hopes of getting one of the fishermen's cobbles. As he was approaching the bridge he learned that the last of the three arches had fallen the instant before; and when he got to the brink, the waters were sweeping on as if it had never been, making the rocks and houses vibrate with a distinct and tremulous motion. The current was playing principally against the southern approach of the bridge, and soon the usually dry arch, at its further end, burst with a loud report; its fragments, mixed with water, being blown into the air as if by gunpowder. The boats had all been swept away, and the fishermen's houses were already one mass of ruin. The centre of the main stream was hurried on at an elevation many feet higher than the rest of the surrounding sea of waters; the mighty rush of which displayed its power in the ruin it p. 55occasioned. Magnificent trees, with all their branches, were dashing and rending against the rock, and the roaring and crashing sound that prevailed was absolutely deafening.

As there was no chance of getting a boat the Doctor returned with difficulty to the house, his mare swimming a great part of the way. On again looking through the telescope at poor Kerr and his family, they were seen huddled together on a spot of ground a few feet square, some forty or fifty yards below their inundated dwelling. [55] He was sometimes standing and sometimes sitting on a small cask, and, as the beholders fancied, watching with intense anxiety the progress of the flood, and trembling for every large tree that it brought sweeping past them. His wife, covered with a blanket, sat shivering on a bit of a log, one child in her lap, and a girl of about seventeen, and a boy of about twelve years of age, leaning against her side. A bottle and a glass on the ground near the man gave the spectators, as it had doubtless given him, some degree of comfort. Above a score of sheep were standing around, or wading, or swimming in the p. 56shallows. Three cows and a small horse picking at a broken rick of straw that seemed to be half afloat, were also grouped with the family. Dreading that they must all be swept off, if not soon relieved, the gentlemen hastened to the offices, and looked anxiously out from the top of the tower for a boat. At last they had the satisfaction to see one launched from the garden at Earnhill, about a mile below. The boat had been conveyed by a pair of horses, and had only just arrived. It was nobly manned by three volunteers, and they proceeded at once to the rescue of a family who were in a most perilous situation in the island opposite to Earnhill. The gentlemen on the tower watched the motions of this boat with the liveliest interest. They saw it tugging up till it was hid from them by the wood. Again it was seen beyond, and soon it dashed into the main stream and disappeared again behind the wood, with a velocity so fearful that they concluded it was lost. But in a moment it again showed itself, and the brave fellows were seen plying their oars across the submerged island of Earnhill, making for John Smith's cottage; the thatch and a small part of the side walls of which were p. 57visible above the water. The poor inmates were dragged out of the windows from under the water, having been obliged to duck within ere they could effect their escape. The boat then swept down the stream towards a place called 'The Lakes,' where John Smith, his wife, and her mother were safely landed.

The boat was next conveyed by the horses to a point from which it was launched for the rescue of the Kerrs. Having pulled up as far as they could in the still water, they approached the desperate current, and fearlessly dashed into its tumultuous waves. For a moment the spectators were in the most anxious doubt as to the result; for, though none could pull a stronger oar, yet the boat in crossing a distance equal to its own length was swept down 200 yards. Ten yards more would have dashed them to atoms on the lower stone wall. But they were now in comparatively quiet water; and availing themselves of this, they pulled up again to the park, in the space between two currents, and passed, with a little less difficulty, though in the same manner, the second and third streams, and at length reached the houses. The spectators gave them three hearty cheers. p. 58By this time the Kerrs had been left scarcely three feet of ground to stand on, under the back wall of the houses. A pleasing sight it was to see the boat touch that tiny strand, and the despairing family taken on board. How anxiously did the spectators watch every motion of the little boat, that was now so crowded as very much to impede the rowers. They crossed the first two streams, and finally drew up for the last and dreadful trial. There the frail bark was again whirled down; and notwithstanding all their exertions, the stern just touched the wall. The prow however was in stiller water; one desperate pull,—she sprang forward in safety, and a few more strokes of the oar landed the poor people amongst fifty or sixty of their assembled friends. After mutual greetings and embraces, and many tears of gratitude, old Kerr related his simple story. "Seeing their retreat cut off by the flood, they attempted to wade ashore. But the nearer the shore, the deeper and more powerful was the current. The moment was awful. The torrent increased on all sides, and night, dark night, was spread over them. The stream began to be too deep for the niece, a girl of twelve years of age,—she lost heart and began to p. 59sink. At this alarming crisis Kerr seized the trembling girl, and placed her on his back, and shoulder to shoulder with his wife, he providentially, but with the greatest difficulty, regained his own house. Between eight and nine o'clock he groped his way, and led his wife and niece up into the garret. He could not tell how long they remained there, but supposed it might be till about two o'clock next morning, when the roof began to fail. To avoid being crushed to death, he worked anxiously till he drove down the partition separating them from the adjoining house. Fortunately for him it was composed of wood and clay, and a partial failure he found in it very much facilitated his operations. Having made their way good, they remained there till about eight o'clock in the morning, when the strength of the water without became so great that it bent inwards the bolt of the lock of the house-door, till it had no greater hold of the staple than the eighth-part of an inch. Aware, that if the door should give way the back wall of the house would be swept down by the rush of the water inwards, and that they would be crushed to atoms, he rummaged the garret and fortunately found a bit of board and a few nails; p. 60and standing on the stairs, he placed one end of it against the door and the other on the hatch, forming the entrance to the garret, and so nailed it firmly down. At last the roof of the second house began to crack over their heads, and Kerr forced a way for himself and his companions through the thatch as has been already told."

Poor Funns and his family were not yet rescued from their little island; and the boat was declared to be too small and weak for so desperate a voyage. It was therefore determined to row to a spot where a larger boat was moored. To effect this, they were compelled to act precisely as they had done in proceeding to rescue the Kerrs. But unfortunately, on entering the third stream, they permitted the boat to glide down with it, in the hope that it would carry them in safety through the gate of the field, and across the road into that beyond it. In this, however, they were mistaken, and the boat was swamped. Fortunately for them, they were carried into smooth water, and by wading shoulder deep they reached the large boat.

Having secured the small boat, they attempted to drag the large one through the gateway against the stream; but it soon filled with water and p. 61swamped, and, in spite of all their exertions, they found it impossible to get it up. The small boat was now all they had to trust to, and this was next caught by the strong stream and overwhelmed in a moment; and had not the men, most providentially, caught and clung to a haycock that happened to be floating past, they must have been lost. They were carried along till it stuck on some young alder trees, when each of them grasped a bough, and the haycock sailed away, leaving them among the weak and brittle branches. They had been here about two hours, when one of the men being unable to hold on longer by the boughs, let himself gently down into the water with the hope of finding bottom; when, to his surprise, he found that the small boat had actually drifted to the root of the very tree to which they had been carried. Some salmon nets and ropes had also, by the strangest accident, been lodged there. The man contrived to pull up one of these with his foot, and making a noose, and slipping it on his great toe, he descended once more, and managed to fix the rope round the stern of the boat, which was then safely hauled up, the oars, being fixed to the side, being also saved. The boat was returned to Mr. p. 62Suter's and fresh manned, when it proceeded to a house occupied by a family of the name of Cumin, consisting of an old couple, their daughter, and grandson. By the time they reached the cottage, its western side was entirely gone, and the boat was pushed in at the gap. Not a sound was heard within, and they suspected that all were drowned; but, on looking through a hole in a p. 63partition, they discovered the unhappy inmates roosted, like fowls, on the beams of the roof. They were, one by one, transferred safely to the boat, half dead with cold; and melancholy to relate, the old man's mind, being too much enfeebled to withstand the agonizing apprehensions he had suffered, was now utterly deranged.

The poor Funns' were still the last to be relieved. They and their cattle were clustered on their little speck of land; and the poor quadrupeds, being chilled by standing so long in the water, were continually pressing inwards on them. It was between six and seven o'clock, the weather was clearer, and the waters were subsiding. The task being the most difficult of all, none but the most skilful rowers were allowed to undertake it. One wide inundation stretched from Monro's house to the tiny spot where Funns and his family were; and five tremendously tumultuous streams raged through it with elevated waves. The moment they dashed into the first of them they were whirled down for a great way; but having once got through it, they pulled up in the quieter water beyond, to prepare for the next; and in doing so, Sergeant Grant stood in the prow, and with a long p. 64rope, the end of which was fixed to the boat, and wherever he thought he had footing, he sprang out and dragged them up. The rest followed his example, and in this way they were enabled to start afresh with a sufficient advantage, and they crossed all the outer streams in the same manner. The last they encountered, being towards the middle of the flood, was fearful, and carried them very far down. But Funns himself, overjoyed to behold them, waded towards them, and gave them his best help to drag up the boat again. Glad was he to see his wife and children safely set in the boat. The perils of their return were not few; but they were at length happily landed.

These examples will suffice to show the nature and extent of the great floods of Moray. The inundation covered a space of something more than twenty miles in the Plain of Forres, and, as it was expressively remarked by one of the sufferers, "Before these floods was the Garden of Eden and behind them a desolate wilderness." And how often did the beautiful expression of the Psalmist occur to them: "The floods have lifted up, O Lord, the floods have lifted up their voice; the floods lift up their waves. The Lord on high is mightier than the noise of many waters; yea, than the mighty waves of the sea." Ps. xciii. 3, 4.

But it is not in Scotland alone that the terrors of the floods are experienced. All rivers which rise in high and cold regions, and pass into warm lowlands, are naturally very liable to overflow their bounds. A remarkable example is afforded by the river Rhone, which rises in the glaciers of Switzerland; and, after passing through the lake of Geneva, descends into the south-eastern departments of France,—a very level district, where the climate is mild and genial. Rapid meltings of the ice in Switzerland, or heavy falls of rain or snow in that country, greatly affect this river; and never, perhaps, were the effects more dreadful than in the inundations of 1840. At Lyons, where the Rhone joins the Saone, the most lamentable scenes took place. Not only were the whole of the low-lying lands in the vicinity of the city completely desolated, hundreds of houses overturned, and many cattle swept away, but the waters reached the city itself, bursting into the gas conduits, and thus leaving the people in darkness, and rising to a great height in the streets. The destruction of property, both in-doors and out-of-doors, was immense, and the loss of life appalling. Charitable people and public servants went about in boats laden with provisions, which were sent, at the expense of the magistrates and clergy, to the starving families pent up in their several abodes, where many of them remained in total darkness by night, and under hourly expectation that the foundations of their houses would give way beneath the rushing waters. In fact, numbers of houses, and even whole streets, were in this way sapped and overturned. Some of the people had fled to the heights near the city, at the first rising of the waters, but there they were reduced to the greatest extremities for want of food, and signal shots were heard from them continually. This miserable state of things lasted from the beginning of November until the 20th or 21st of the same month. At the same time the Rhone appeared like a succession of immense lakes from Lyons to Avignon, and from Avignon to the sea. A letter from Nismes, a little to the west of Avignon, thus described the scene:—

> "As far as the view extends we perceive but one sheet of water, in the midst of which appear the tops of trees and houses, with the miserable inhabitants perched upon them. At Valabrhgue, an island on the Rhone, they have hung out a black banner from the church-yard, nearly two thousand persons being assembled in that spot, which is on an elevation. Steam-boats are attempting to carry bread to Valabrhgue, and other similarly situated places, but can scarcely effect it from the inequality of the ground. For ten days the rains have never ceased. The space covered by the waters near Avignon is calculated at about thirty-six leagues in length and sixty leagues in breadth. Human bodies are seen passing continually on the waters."

From the 10th to the 20th of November the Rhone fell several inches each day, but always rose again somewhat during the night. It began permanently to decline on the 20th, and in a few days the streets were exposed to view, with about a foot of mud on them. The loss of life and property, through this calamity, are almost incalculable.

A still grander display of the power and extent of inundations is afforded by the American rivers. The mighty waters of the Mississippi, (a river, whose course extends for several thousand miles,) when swelled, and overflowing their banks, present a wonderful spectacle. Unlike the mountain-torrents, and small rivers, of other parts of the world, the Mississippi rises slowly, continuing for several weeks to increase at the rate of about an inch in a day. When at its height, it undergoes little change for some days, and after this subsides as slowly as it rose. A flood generally lasts from four to six weeks, though it sometimes extends to two months. The American naturalist, Audubon, has given a striking account of the rush of waters overspreading the land when once this mighty river has begun to overflow its banks:—

> "No sooner has the water reached the upper part of the banks, than it rushes out, and overspreads the whole of the neighbouring swamps, presenting an ocean overgrown with stupendous forest trees. So sudden is the calamity that every individual, whether man or beast, has to exert his utmost ingenuity to enable him to escape from the dreaded element. The Indian quickly removes to the hills of the interior, the cattle and game swim to the different strips of land that remain uncovered in the midst of the flood, or attempt to force their way through the waters until they perish from p. 69fatigue. Along the banks of the river the inhabitants have rafts ready-made, on which they remove themselves, their cattle, and their provisions, and which they then fasten with ropes or grape-vines to the larger trees, while they contemplate the melancholy spectacle presented by the current, as it carries off their houses and woodyards piece by piece. Some, who have nothing to lose, and are usually known by the name of Squatters, take this opportunity of traversing the woods in canoes, for the purpose of procuring game, and particularly the skins of animals, such as the deer and bear, which may be converted into money. They resort to the low ridges surrounded by the waters, and destroy thousands of deer, merely for their skins, leaving the flesh to putrify.
>
> "The river itself, rolling its swollen waters along, presents a spectacle of the most imposing nature. Although no large vessel, unless propelled by steam, can now make its way against the current, it

is seen covered by boats laden with produce, which, running out from all the smaller streams, float silently towards the city of New Orleans, their owners, meanwhile, not very well assured of finding a landing-place even there. p. 70The water is covered with yellow foam and pumice, the latter having floated from the rocky mountains of the north-west. The eddies are larger and more powerful than ever. Here and there tracts of forest are observed undermined, the trees gradually giving way, and falling into the stream. Cattle, horses, bears, and deer are seen at times attempting to swim across the impetuous mass of foaming and boiling water; whilst, here and there, a vulture or an eagle is observed perched on a bloated carcass, tearing it up in pieces, as regardless of the flood, as on former occasions it would have been of the numerous sawyers and planters with which the surface of the river is covered when the water is low. Even the steamer is frequently distressed. The numberless trees and logs that float along, break its paddles, and retard its progress. Besides it is on such occasions difficult to procure fuel to maintain its fires."

In certain parts, the shores of the Mississippi are protected by artificial barriers called Levies. In such places, during a flood, the whole population of the district is engaged in strengthening these barriers, each proprietor being in great alarm lest a crevasse should open and let in the p. 71waters upon his fields. In spite of all exertions this disaster generally happens: the torrent rushes impetuously over the plantations, and lays waste the most luxuriant crops.

The mighty changes effected by the inundations of the Mississippi are little known until the waters begin to subside. Large streams are then found to exist where none had formerly been. These are called by navigators *short cuts*, and some of them are so considerable as to interfere with the navigation of the Mississippi. Large sand-banks are also completely removed by the impetuous whirl of the waters, and are deposited in other places. Some appear quite new to the navigator, who has to mark their situation and bearings in his log-book. Trees on the margin of the river have either disappeared, or are tottering and bending over the stream preparatory to their fall. The earth is everywhere covered by a deep deposit of muddy loam, which, in drying, splits into deep and narrow chasms, forming a sort of network, from which, in warm weather, noxious exhalations rise, filling the atmosphere with a dense fog. The Squatter, shouldering his rifle, makes his way through the morass in search of his p. 72lost stock, to drive the survivors home and save the skins of the drowned. New fences have everywhere to be formed, and new houses erected; to save which from a like disaster, the settler places them on a raised platform, supported by pillars made of the trunks of trees. "The lands must be ploughed anew; and if the season is not too far advanced, a crop of corn and potatoes may yet be raised. But the rich prospects of the planter are blasted. The traveller is impeded in his journey, the creeks and smaller streams having broken up their banks in a degree proportionate to their size. A bank of sand, which seems firm and secure, suddenly gives way beneath the traveller's horse, and the next moment the animal has sunk in the quicksand, either to the chest in front, or to the crupper behind, leaving its master in a situation not to be envied."

p. 74

CHAPTER III.

various forms op clouds—the cirrus, or curl-cloud—the cumulus, or stacken-cloud—the stratus, or fall-cloud—the cirro-cumulus, or sonder-cloud—the cirro-stratus, or wane-cloud—the cumulo-stratus, or twain-cloud—the nimbus, or rain-cloud—arrangement of rain-clouds—appearances of a distant shower—scud—cause of rain—formation of clouds—mists—heights of clouds—appearance of the sky above the clouds.

Many persons are apt to suppose that the clouds are among the most fitful and irregular appearances in the world; fleeting and unstable in their nature, uncertain in their forms, apparently subject to no fixed laws, and obedient neither to times nor seasons. Attentive observers, however, have proved that the beauty and harmony which are everywhere found to prevail in nature when rightly understood, can also be traced, even in the clouds. Although very much still remains to be discovered respecting them, yet it is found that, like all the other natural productions, they admit of being arranged and classified. So obvious was this to persons whose interest it is to observe the weather, that, long before scientific men had studied the subject, country people had noticed the different forms of clouds, and had learned to distinguish them by different names.

The first scientific man who made the clouds the object of his particular study, was Luke Howard, who, from an attentive consideration of their forms and appearances, found that they might all be arranged under three simple or primary forms, namely:—

1. The *Cirrus*—so called from its resemblance to a *curled lock of hair*. (Figures, 1, 2; page 77.)

2. The *Cumulus*, from the *heaped* appearance presented by the convex masses which form this cloud. (Figure 7.)

3. The *Stratus*, from its spreading out horizontally in a continuous layer, and increasing from below. (Figure 10.)

These three primary forms are subject to four modifications:—

The first is the *Cirro-cumulus*, consisting of small roundish and well-defined masses, in close horizontal arrangement. (Figure 3.)

The second is the *Cirro-Stratus*, and the masses p. 78which compose it are small and rounded, but thinned off towards a part, or towards the whole of their circumference. They are sometimes separate, and sometimes in groups. (Figures 4, 5, 6.)

The third is the *Cumulo-Stratus*, which is made up of the cirro-stratus blended with the cumulus. (Figure 8.)

The fourth is the *Cumulo-Cirro-Stratus*, or *Nimbus*. This is the true *rain-cloud*, or system of clouds from which rain is falling. (Figure 9.)

The term *modification* applies to the structure or manner in which a given mass of cloud is made up, and not to its precise form or size, which in most clouds varies every instant. Mr. Howard remarks, that it may be at first difficult to distinguish one modification from another, or to trace the narrow limits which sometimes separate the different modifications; but a moderate acquaintance with the subject will soon enable any one to point out the various forms, and to a great extent to judge of the state of the weather by them. In order, therefore, to assist the reader in gaining a certain amount of knowledge on this interesting subject, it may be useful to state more fully the various phenomena of the different forms of clouds already enumerated.

p. 79

The Cirrus occurs in very great variety, and in some states of the air is constantly changing. It is the first cloud that appears in serene weather, and is always at a great height. The first traces of the p.

80cirrus are some fine whitish threads, delicately-pencilled on a clear blue sky; and as they increase in length others frequently appear at the sides, until numerous branches are formed, extending in all directions. Sometimes these lines cross each other and form a sort of delicate net-work. In dry weather the cirrus is sharp, defined, and fibrous in texture, the lines vanishing off in fine points. When the air is damp this cloud may be seen in the intervals of rain, but is not well defined, and the lines are much less fibrous. Such cirri as these often grow into other varieties of cloud, and are frequently followed by rain.

The cirrus may last a few minutes only, or continue for hours. Its duration is shortest when near other clouds. Although it appears to be stationary, it has some connexion with the motions of the atmosphere; for whenever, in fair weather, light variable breezes prevail, cirri are generally present. When they appear in wet weather, they quickly pass into the cirro-stratus.

According to Dalton, these clouds are from three to five miles above the earth's surface. When viewed from the summits of the highest mountains they appear as distant as from the p. 81plains. Another proof of their great height is, their continuing to be tinged by the sun's rays in the evening twilight with the most vivid colours, while the denser clouds are in the deepest shade.

The cirrus appears to be stationary; but, on comparison with a fixed object, it will sometimes be found to make considerable progress.

THE CUMULUS, OR STACKEN-CLOUD.

> "And now the mists from earth are clouds in heaven:
> Clouds, slowly castellating in a calm
> Sublimer than a storm; while brighter breathes
> O'er the whole firmament the breadth of blue,
> Because of that excessive purity
> Of all those hanging snow-white palaces,
> A gentle contrast, but with power divine."

The Cumulus is a day cloud; it usually has a dense, compact appearance, and moves with the wind. In the latter part of a clear morning a small irregular spot appears suddenly at a moderate elevation. This is the nucleus or commencement of the cloud, the upper part of which soon becomes rounded and well defined, while the lower forms an irregular straight line. The cloud evidently increases in size on the convex surface, p. 82one heap succeeding another, until a pile of cloud is raised or *stacked* into one large and elevated mass, or *stacken-cloud*, of stupendous magnitude and beauty, disclosing mountain summits tipped with the brightest silver; the whole floating along with its point to the sky, while the lower surface continues parallel with the horizon.

p. 83When several cumuli are present, they are separated by distances proportioned to their size: the smaller cumuli crowding the sky, while the larger ones are further apart. But the bases always range in the same line; and the increase of each cloud keeps pace with that of its neighbour, the intervening spaces remaining clear.

The cumulus often attains its greatest size early in the afternoon, when the heat of the day is most felt. As the sun declines, this cloud gradually decreases, retaining, however, its characteristic form till towards sunset, when it is, more or less, hastily broken up and disappears, leaving the sky clear as in the early part of the morning. Its tints are often vivid, and pass one into the other in a most pleasing manner, during this last hour of its existence.

This cloud accompanies and foretells fine weather. In changeable weather it sometimes evaporates almost as soon as it is formed; or it appears suddenly, and then soon passes off to some other modification.

In fair weather this cloud has a moderate elevation and extent, and a well-defined rounded surface. Before rain it increases more rapidly than p. 84at other times, and appears lower in the atmosphere, with its surface full of loose fleeces.

The formation of large cumuli to leeward, in a strong wind, indicates the approach of a calm with rain. When they do not disappear or subside about sun-set, but continue to rise, thunder is to be expected in the night.

Independently of the beauty and magnificence which this description of cloud adds to the face of nature, it serves to screen the earth from the direct rays of the sun; by its multiplied reflections to diffuse and, as it were, economise the light; and also to convey immense stores of vapour from the place of its origin to a region in which moisture may be wanted.

THE STRATUS, OR FALL-CLOUD.

As the Cumulus belongs to the day, so does the Stratus to the night. It is the lowest of all the clouds, and actually rests upon the earth, or the surface of water. It is of variable extent and thickness, and is called *Stratus, a bed* or *covering*. It is generally formed by the *sinking* of vapour in the atmosphere, and on this account has been p. 85called *Fall-cloud*. It comprehends all those level, creeping mists, which, in calm evenings, spread like an inundation from the valleys, lakes, and rivers, to the higher ground. [85] But on the return of the sun the beautiful level surface of this p. 86cloud begins to put on the appearance of cumulus, the whole, at the same time, rising from the ground like a magnificent curtain. As the cloud ascends, it is broken up and evaporates or passes off with the morning breeze. The stratus has long been regarded as the harbinger of fine weather; and, indeed, there are few days in the year more serene than those whose morning breaks out through a stratus.

THE CIRRO-CUMULUS, OR SONDER-CLOUD.

The cirrus having continued for some time increasing or stationary, usually passes either to the cirro-cumulus or to the cirro-stratus, at the same time descending to a lower station in the atmosphere.

The Cirro-cumulus is formed from a cirrus, or a number of small separate cirri, passing into roundish masses, in which the extent of the cirrus is no longer to be seen. This change takes place either throughout the whole mass at once, or progressively from one extremity to the other. In either case the same effect is produced on a number of neighbouring cirri at the same time, and in p. 87the same order. It appears, in some instances, to be hastened by the approach of other clouds.

The cirro-cumulus forms a very beautiful sky, exhibiting sometimes numerous distinct beds of small connected clouds floating at different heights. It is frequent in summer, and accompanies warm, p. 88dry weather. On a fine summer's evening the small masses which compose this cloud, are often well defined, and lying quite *asunder*, or separate from one another; and on this account the term *sonder-cloud* has been applied to this modification. The whole sky is sometimes covered with these small masses. They are occasionally, and more sparingly, seen in the intervals of showers, and in winter.

Bloomfield, in the following beautiful lines, has noticed the appearance of the sonder-cloud: —

> "For yet above these wafted clouds are seen
> (In a remoter sky still more serene)

> Others, detach'd in ranges through the air,
> Spotless as snow, and countless as they're fair;
> Scatter'd immensely wide from east to west,
> The beauteous semblance of a flock at rest:
> These, to the raptur'd mind, aloud proclaim
> The mighty Shepherd's everlasting name."

This cloud may either evaporate or disappear, or it may pass to the cirrus, or sink lower and become a cirro-stratus. In stormy weather, before thunder, a cirro-cumulus often appears, composed of very dense and compact round bodies, in very close arrangement. When accompanied by the cumulo-stratus, it is a sure indication of a coming storm.

THE CIRRO-STRATUS, OR WANE-CLOUD.

This cloud appears to be formed from the fibres of the cirrus sinking into a horizontal position, at the same time that they approach each other sideways. This cloud is to be distinguished by its flatness and great horizontal extension, in proportion to its height; a character which it always retains, under all its various forms. As this cloud is generally changing its figure, and slowly sinking, it has been called the *wane-cloud*. A collection of these clouds, when seen in the distance, frequently give the idea of shoals of fish. Sometimes the whole sky is so mottled with them, as to obtain for it the name of the *mackerel-back sky*, from its great resemblance to the back of that fish. Sometimes they assume an arrangement like discs piled obliquely on each other. But in this, as in other instances, the structure must be attended to rather than the form, for this varies much, presenting, at times, the appearance of parallel bars or interwoven streaks, like the grain of polished wood. It is thick in the middle and thinned off towards the edge.

These clouds precede wind and rain. The near or distant approach of a storm may often be judged of from their greater or less abundance and duration. They are almost always to be seen in the p. 91intervals of storms. Sometimes the cirro-stratus, and the cirro-cumulus, appear together in the sky, and even alternate with each other in the same cloud, presenting many curious changes; and a judgment may be formed of the weather likely to ensue, by observing which prevails at last.

The cirro-stratus most frequently forms the solar and lunar halo. Hence the reason of the prognostics of bad weather commonly drawn from the appearance of halos.

THE CUMULO-STRATUS, OR TWAIN-CLOUD.

This is a blending of two kinds of cloud (hence the name of *twain-cloud*,) and it often presents a grand and beautiful appearance, being a collection of large fleecy clouds overhanging a flat stratum or base. When a cumulus increases rapidly a cumulo-stratus frequently forms around its summit, resting thereon as on a mountain, while the former cloud continues to be seen, in some degree, through it. This state of things does not continue long. The cumulo-stratus speedily becomes denser and spreads, while the upper part of the cumulus extends likewise, and passes into it, the base continuing as it p. 92was. A large, lofty, dense cloud is thus formed which may be compared to a mushroom with a very thick, short stem. The cumulo-stratus, when well formed and seen singly, and in profile, is quite as beautiful an object as the cumulus. Mr. Howard has occasionally seen specimens constructed almost p. 93as finely as a Corinthian capital; the summit throwing a well-defined shadow upon the parts beneath. It is sometimes built up to a great height. The finest examples occur between the first appearance of the fleecy cumuli and the commencement of rain, while the lower atmosphere is comparatively dry, and during the approach of thunder storms. The appearance of the cumulo-stratus, among ranges of hills, presents some interesting phenomena. It appears like a curtain dropping among them and enveloping their summits; the hills reminding the spectator of the massy Egyptian columns which support the flat-roofed temples of Thebes. But when a whole sky is crowded with these clouds, and the cumulus rises behind them, and is seen through the interstices, the whole, as it passes off in the distant horizon, presents to the fancy mountains covered with snow, intersected with darker ridges, lakes of water, rocks and towers. Shakspeare seems to have referred to this modification in the well-known lines:—

> "Sometimes we see a cloud that's dragonish;
> A vapour, sometimes, like a bear or lion,
> A towered citadel, a pendent rock,
> A forked mountain, a blue promontory,
> p. 94With trees upon't that nod unto the world,
> And mock our eyes with air.—
> That which is now a horse, even with a thought
> The rack dislimns, and makes it indistinct
> As water is in water.

The *distinct* cumulo-stratus is formed in the interval between the first appearance of the fleecy *cumulus* and the commencement of rain, while the lower atmosphere is yet dry; also during the approach of thunder storms when it has frequently a reddish appearance. Its *indistinct* appearance is chiefly in the longer or shorter intervals of showers of rain, snow, or hail.

THE CUMULO-CIRRO-STRATUS; NIMBUS OR RAIN-CLOUD.

Clouds, in any one of the preceding forms, at the same degree of elevation, or two or more of these forms at different elevations, may increase and become so dense as completely to obscure the sky; this, to an inexperienced observer, would seem to indicate the speedy commencement of rain. But Mr. Howard is of opinion that clouds, while p. 95in any of the states above described, never let fall rain.

Before rain the clouds always undergo a change of appearance, sufficiently remarkable to give them a distinct character. This appearance, when the rain happens overhead, is but imperfectly seen; but from the observations of akronauts, it appears that whenever a fall of rain occurs, and the sky is at the same time entirely overcast with clouds, there will be found to exist another stratum of clouds at a certain elevation above the former. So, also, when the sky is entirely overcast and rain is altogether or generally absent, the akronaut, upon traversing the canopy immediately above him, is sure to enter upon an upper hemisphere either perfectly cloudless or nearly so. These remarks were, we believe, first made by Mr. M. Mason, and he states that they have been verified during many hundred ascents.

In October, 1837, two ascents were made by Mr. Mason, which well illustrate what has been said. On the 12th, "the sky was completely overspread with clouds, and torrents of rain fell incessantly during the whole of the day. Upon quitting the earth, the balloon was almost immediately p. 96enveloped in the clouds, through which it continued to work its way upwards for a few seconds. Upon emerging at the other side of this dense canopy, a vacant space, of some thousand feet in breadth, intervened, above which lay another stratum of a similar form and observing a similar character. As the rain, however, still continued to pour from this second layer of clouds, to preserve the correctness of the observation, a third layer should, by right, have existed at a still further elevation; which, accordingly, proved to be the case. On the subsequent occasion of the ascent of the same balloon, (October 17th,) an exactly similar condition of the atmosphere, with respect to clouds, prevailed; unaccompanied, however, with the slightest appearance of rain. No sooner had the balloon passed the layer of clouds immediately above the surface of the earth, than, as was anticipated, not a single cloud was to be found in the firmament beyond; an unbroken expanse of clear blue sky everywhere embracing the frothy plain that completely intercepted all view of the world beneath."

Mr. Howard had not the advantages of a balloon to assist his observations. He has noticed that during rain and before the arrival of the denser p. 97and lower clouds, or through their interstices, there exists, at a greater height, a thin light veil or a hazy appearance. When this has considerably increased, the lower clouds are seen to spread till they unite in all points and form one uniform sheet. The rain then commences, and the lower clouds arriving from the windward, move under this sheet and are successively lost in it. When the latter cease to arrive, or when the sheet breaks, letting through the sun-beams, every one's experience teaches him to expect that the rain will abate or leave off.

But there often follows an immediate and great addition to the quantity of cloud. At the same time the darkness becomes less, because the arrangement, which now returns, gives free passage to the rays of light; the lower broken clouds rise into cumuli, and the upper sheets put on the various forms of the cumulo-stratus, sometimes passing to the cirro-cumulus.

The various phenomena of the rain-cloud are best seen in a distant shower. If the cumulus be the only cloud at first visible, its upper part is seen to become tufted with cirri. Several adjacent clouds also approach and unite at its side. The cirri p. 98increase, extending upwards and sideways, after which the shower is seen to commence. At other times, the cirro-stratus is first formed above the cumulus, and their sudden union is attended with the production of cirri and rain. In either case the cirri spring up in proportion to the quantity of rain falling, and give the cloud a character by which it is easily known at great distances, and which has long been called by the name of *nimbus*.

When one of these arrives hastily with the wind, it brings but little rain, and frequently some hail or driven snow.

Since rain may be produced and continue to fall from the slightest obscuration of the sky by the nimbus, while a cumulus or a cumulo-stratus, of a very dark and threatening aspect, passes on without discharging any until some change of state takes place; it would seem as if nature had destined the latter as reservoirs, in which water is collected from extensive regions of the air for occasionally irrigating particular spots in dry seasons; and by means of which it is arrested, at times, in its descent in wet ones.

Although the nimbus is one of the least beautiful of clouds, it is, nevertheless, now and then adorned p. 99by the splendid colouring of the rainbow, which can only be seen in perfection when the dark surface of this cloud forms for it a background.

The small ragged clouds which are sometimes seen sailing rapidly through the air, are called *scud*. They consist of portions of a raincloud, probably broken up by the wind, and are dark or light according as the sun shines upon them. They are the usual harbingers of rain, and, as such, are called by various names, such as *messengers, carriers,* and *water-waggons*.

In attempting to explain the production of clouds and rain, it is necessary to observe that the subject is beset with difficulties—the discussion of which does not belong to this little volume; but the following notice of Dr. Hutton's theory may not be out of place.

It has been already stated, that the air supplies itself with moisture from the surface of the waters of the earth. This it continues to do at all temperatures, until it is so charged with vapour that it cannot contain any more. The air is then said to be *saturated*. Now, the quantity of moisture which a given bulk of air can contain, depends entirely p. 100upon the temperature of the air for the time being. The higher the temperature of the air the greater will be the quantity of vapour contained in it; and, although it may be perfectly invisible to the eye, on account of the elasticity which the heat imparts to it, yet it can easily be made visible by subtracting a portion of the heat. If, for example, a glass of cold water be suddenly brought into a warm room, moisture from the air will be condensed upon the outside of the glass in the form of dew. A similar change is supposed to take place when two currents of air having different temperatures, but both saturated with vapour, are mingled together; an excess of vapour is set free, which forms a cloud or falls down as rain. If the currents continue to mingle uniformly, "the clouds soon spread in all directions, so as to occupy the whole horizon; while the additional moisture, incessantly brought by the warmer current, keeps up a constant supply for condensation, and produces a great and continued deposition of moisture in the form of rain. By degrees, the currents completely intermingle, and acquire a uniform temperature; condensation then ceases; the clouds are re-dissolved; and the whole face of nature, after being cooled and refreshed p. 101by the necessary rain, is again enlivened by the sunshine, thus rendered still more agreeable by its contrast with the previous gloom."

If the cloud, produced by the mingling of two differently heated currents of moist air, happen to form in the upper regions of the sky, it may be heavier than its own bulk of air, and will consequently begin to sink. Should the atmosphere near the earth be less dense than the cloud, the latter will continue to descend till it touches the ground, where it forms a mist. If the vapour has been condensed rapidly and abundantly, the watery particles will form rain, hail, or snow, according to the temperature of the air through which they pass. But it may happen that the cloud, in descending, arrives in a warmer region than that in which it was formed: in this case, the condensed moisture may again become vapour, and ascend again to a region where condensation may again take place.

Mr. Daniell's explanation of the formation of rain differs from the above in some of its particulars, which are not sufficiently elementary to be given here; but it may be instructive to give a few of Mr. Howard's illustrations respecting the p. 102formation of the various clouds. If hot water be exposed to cool air, it *steams*—that is, the vapour given off from the surface is condensed in mixing with the air; and the water thus produced appears in visible particles, the heat of the vapour passing into the air. This effect may be seen about sunrise, in summer, on the surface of ponds warmed by the sun of the previous day, and also with water newly pumped from a well. But the small cloud formed in these instances usually disappears almost as soon as formed, the air being too dry to allow it to remain. But in the wide regions of the atmosphere the case is different, on account of the vast supply of vapour, and the ascent and descent of the cloud to regions which allow it to remain tolerably permanent. In the fine evenings of autumn, and occasionally at other seasons, mists appear suddenly in the valleys, gradually filling these low places, and even rising to a certain height, forming a foggy atmosphere for the following day. These collections of visible vapour resting on the earth, and often cut off so as to form a level surface above, so nearly resemble a sheet of water, as to have been occasionally mistaken for an inundation, the occurrence of the previous night. Such is the p. 103origin and appearance of the *stratus*: it constitutes the fog of the morning, and sometimes, as at the approach of a long frost, occupies the lower atmosphere for several days. But the sun, we will suppose, has broken through and dissipated this obscurity, and cleared the lower air. On looking up to the blue sky, we see some few spots showing the first formation of a cloud there: these little collections increase in number, and become clouds, heaped, as it were, on a level base, and presenting their rounded forms upwards; in which state they are carried along in the breeze, remaining distinct from each other in the sky. This is the *cumulus*, or *heap*.

By and by, if the clouds continue to form, and enough vapour is supplied from above, these heaps are seen to grow over their base like a mushroom or cauliflower. Perhaps a flat top is seen forming separately, and this afterwards joins the simple heap of cloud; or the flat forms and the heaps become mixed irregularly among each other, occupying the spaces everywhere, till the sky becomes overcast, and presents the usual appearance of dense clouds. This is the *cumulo-stratus*, or *heaped and flat cloud*. It is not productive of rain, and it p. 104forms, both in summer and in winter, the common scenery of a full sky.

On examining minutely the higher regions of the air, especially after the sky has been clear for some time, the spectator will probably see the cirrus descending from above in the form of *threads* or *locks* and *feathers*, which go on increasing until they fill the sky. They are more commonly seen above the two former kinds, which float upon the clear air below. On continuing to watch the cirri, they will be seen to pass to the intermediate form of cirro-cumulus, consisting of smaller rounded clouds attached to each other, or simply collected together in a flat aggregate, and forming the mottled or dappled sky.

The cumulo-stratus is more dense and continuous in its structure; thick in the middle, and thinned off towards the edges. Over-head it is a mere bed of haze, more or less dense. In the horizon, when seen sideways, it often resembles shoals of fish, as already noticed; but it is liable to put on the most ragged and patchy appearances, making a very ugly sky.

The nimbus, or rain cloud, is seen to the greatest advantage in profile, in the horizon, and at a great p. 105distance, when it often resembles a lofty tower raised by its greater height to a conspicuous place among the dark threatening clouds, and catching the sun's last rays upon its broad summit and sides. In its nearer approach, it may always be known by being connected below with an obscurity caused by the rain it lets fall, and which reaches down to the horizon.

In ascending from the lower valleys to the tops of lofty mountains, clouds may be traced through six modifications, the cirrus being seen from the loftiest summits, while the other forms are only skirting the sides of the mountains. Mr. Mason remarks, that clouds occasionally lie so low, that before the balloon seems to have entirely quitted the earth, it has been received between their limits, and entirely enveloped within their watery folds. Clouds, on the contrary, are sometimes at such a height, that the balloon either never comes into contact with them at all, or, if it passes through one layer, the akronaut continues to behold another occupying a still remoter region of the skies above.

As a general rule, it is stated that the natural region of clouds is a stratum of the atmosphere p. 106lying between the level of the first thousand feet, and that of one removed about ten thousand feet above it. Of course it is not supposed but that clouds are occasionally found on both sides of the bounds here assigned to them; the mist occupies the lowest valleys, while, on the other hand, long after the akronaut has attained the height of ten thousand feet, some faint indications of clouds may still be seen partially obscuring the dark blue vault above him. As he continues to ascend, the blue of the sky increases in intensity; and should a layer of clouds shut out all view of the earth, "above and all around him extends a firmament dyed in purple of the intensest hue; and from the apparent regularity of the horizontal plane on which it rests, bearing the resemblance of a large inverted bowl of dark blue porcelain standing upon a rich Mosaic floor or tesselated pavement. Ascending still higher, the colour of the sky, especially about the zenith, is to be compared with the deepest shade of Prussian blue."

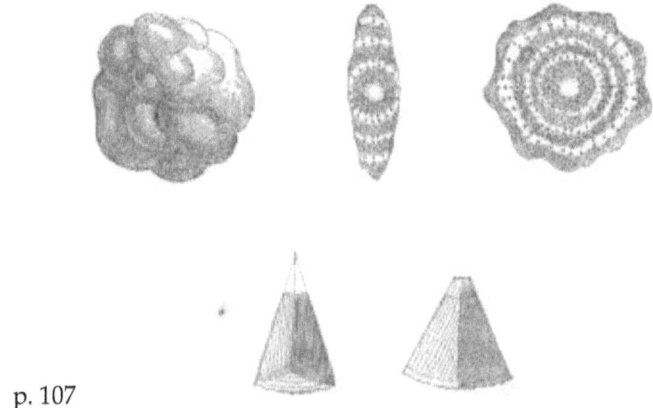

p. 107

CHAPTER IV.

on hail — the hail-storms of france — disastrous effects of hail — the hail-storms of south america — their surprising effects — origin and nature of hail — periodical falls of hail — hail clouds — hailstones — their various forms — extraordinary size of hailstones.

As hail seems to be nothing more than frozen rain, it is necessary to collect a few particulars respecting it in this place.

Great Britain is essentially a rain country; but there are some parts of the world which have obtained the unhappy distinction of being *hail* countries: p. 108such, for example, as some of the most beautiful provinces of France, which are frequently devastated by hail-storms. One of the most tremendous hail-storms on record is that which occurred in that country in July 1788. This fearful storm was ushered in by a dreadful and almost total darkness which suddenly overspread the whole country. In a single hour the whole face of nature was so entirely changed, that no person who had slept during the tempest could have believed himself in the same part of the world when he awoke. Instead of the smiling bloom of summer, and the rich prospects of a forward autumn, which were just before spread over the face of that fertile and beautiful country, it now presented the dreary aspect of an arctic winter. The soil was changed into a morass; the standing corn beaten into a quagmire; the vines were broken to pieces, and their branches bruised in the same manner; the fruit-trees of every kind were demolished, and the hail lay unmelted in heaps like rocks of solid ice. Even the robust forest trees were incapable of withstanding the fury of the tempest; and a large wood of chesnut trees, in particular, was so much damaged, that it presented, after the storm, p. 109little more than bare and naked trunks. The vines were so miserably hacked and battered, that four years were estimated as the shortest period in which they could become again in any degree productive. Of the sixty-six parishes included in the district of Pontoise, forty-three were entirely desolated; while, of the remaining twenty-three, some lost two-thirds, and others above half their harvest.

This storm began in the south, and proceeded in two parallel bands from the south-west to the north-east; the extent of one of them being 175 leagues, and of the other 200; thus traversing nearly the whole length of that great kingdom, and even a portion of the Low countries. The mean breadth of the eastern portion was four leagues, and of the western two: and, what is very remarkable, the interval between the two bands, amounting to five leagues, was deluged with heavy rain. The largest of the hail-stones weighed half a pound each.

The progress of this storm, which was from south to north, was at the rate of 16= leagues an hour; and the velocity of the two bands was precisely the same. The continuance of the hail p. 110was limited to seven or eight minutes, at each of the principal stations marked.

There are instances, however, on record, in which hail has produced even more tremendous results than those above recorded. In some parts of South America hail-stones are sometimes so large and so hard, and fall with such violence, that large animals are killed by them. Mr. Darwin, encamping at the foot of the Sierra Tapalguen, says: — "One of the men had already found thirteen deer lying dead, and I saw their fresh hides. Another of the party, a few minutes after my arrival, brought in seven more. Now I well know that one man without dogs could hardly have killed seven deer in a week. The men believed they had seen about fifteen dead ostriches, (part of one of which we had for dinner;) and they said that several were running about evidently blind in one eye. Numbers of small birds, as ducks, hawks, and partridges, were killed. I saw one of the latter with a black mark on its back, as if it had been struck with a paving-stone. A fence of thistle-stalks round the hovel was nearly broken down; and my informer, putting his head out to see what was the matter, received a severe p. 111cut, and now wears a bandage. The storm was said to have been of limited extent: we certainly saw, from our last night's bivouac, a dense cloud and lightning in this direction. It is marvellous how such strong animals as deer could thus have been killed; but, I have no doubt, from the evidence I have given, that the story is not in the least exaggerated." Dr. Malcolmson informed Mr. Darwin, that he witnessed, in 1831, in India, a hail-storm, which killed numbers of large birds, and much injured the cattle. These hail-stones were flat; one was ten inches in circumference; and another weighed two ounces. They ploughed up a gravel-walk like musket-balls, and passed through glass windows, making round holes, but not cracking them.

There is much in the origin and formation of hail that cannot well be explained. Volta regarded the formation of small flakes of ice, the kernels of future hail-stones, in the month of July, during the hottest hours of the day, as one of the most difficult phenomena in nature to explain. It is difficult to account for the comparative scarcity of hail-showers in winter; as also, for the great size which hailstones are often known to attain.

p. 112It appears from certain resemblances in the descents of rain, snow, and hail, that they have a common origin, their different formations being explained by difference of temperature. Howard has observed a huge nimbus affording hard snowballs and distinct flakes of snow at the same time. Hail and rain are by no means uncommon from the same cloud. The size of a cloud may be such, or clouds may exist in different elevations, which in an upper region produce hail, in a lower region snow, and at a still lower elevation rain. Rain may also form in an upper region of the sky, and descend into a colder stratum of the atmosphere, and be frozen into hail. Hail generally precedes storms of rain.

Change of wind and the action of opposite currents, so necessary for the production of rain, are also frequent during hail-storms. While clouds are agitated with the most rapid motions, rain generally falls in greatest abundance; and if the agitation be very great it generally hails. Before the descent of hail a noise is heard, a particular kind of crackling, which has been compared to the emptying of a bag of walnuts.

The descent of hail in some countries appears p. 113to occur at particular periods. In the central parts of France, Italy, and Spain, it usually hails most abundantly during the warmest hours of the day in spring and summer, and in Europe generally it falls principally during the day; but there are examples recorded of great hail-storms which have taken place during the night. Near the equator, it seldom hails in places situated at a lower level than 350 fathoms, for, although the hail may be formed, the warmth of the regions prevents it from falling in that state.

The appearance of hail clouds seems to be distinguished from other stormy clouds by a very remarkable shadowing. Their edges present a multitude of indentations, and their surfaces disclose here and there immense irregular projections. Arago has seen hail-clouds cover with a thick veil the whole extent of a valley, at a time when the neighbouring hills enjoyed a fine sky and an agreeable temperature.

Hailstones of similar forms are produced at similar levels. They are smaller on the tops of mountains than in the neighbouring plains. If the temperature or the wind alter, the figures of the hailstones become immediately changed. Hailstones p. 114of the form of a six-sided pyramid have been known to change, on the wind changing to the north-east, to convex lenses, so transparent and nicely formed, that they magnified objects without distorting them. Some hailstones are globular, others elongated, and others armed with different points.

In the centres of hailstones small flakes of spungy snow are frequently found, and this usually is the only opaque point in them. Sometimes the surface is covered with dust, like fine flour, and is something between hail and snow. This never falls during summer in southerly countries. In the Andes hailstones from five to seven lines in diameter are sometimes formed of layers of different degrees of transparency, so as to permit rings of ice to be separated from them with a very slight blow. In Orkney, hailstones have fallen as finely polished as marbles, of a greyish white colour, not unlike fragments of light-coloured marble. Hailstones are often so hard and elastic, that those which fall on the stones rebound without breaking to the height of several yards; and they have been known to be projected from a cloud almost horizontally, and with such velocity p. 115as to pierce glass windows with a clear round hole.

On the 7th May, 1822, some remarkable hailstones fell at Bonn, on the Rhine. Their general size was about an inch and a half in diameter, and their weight 300 grains. When picked up whole, which was not always the case, their general outline was elliptical, with a white, or nearly opaque spot in the centre, about which were arranged concentric layers, increasing in transparency to the outside. Some of them exhibited a beautiful star-like and fibrous arrangement, the result of rows of air bubbles dispersed in different radii. The figures at the head of this chapter show the external and internal appearances of these hailstones.

The smaller figures represent pyramidal hail, common in France, and occasionally in Great Britain.

Brown hailstones have been noticed. Humboldt saw hail fall of the colour of blood.

On the 15th July, 1808, Howard noticed, in Gloucestershire, hailstones from three to nine inches in circumference; appearing like fragments of a vast plate of ice which had been broken in its descent to the earth.

p. 116On the 4th June, 1814, Dr. Crookshank noticed, in North America, hailstones of from thirteen to fifteen inches in circumference. They seemed to consist of numerous smaller stones fused together.

On the 24th July, 1818, during a storm in Orkney, Mr. Neill picked up hailstones weighing from four ounces to nearly half a pound.

p. 117

CHAPTER V.

method of measuring the quantity of rain that falls—the rain gauge—methods of observing for rain and snow—effects of elevation on the quantity of rain—difference between the top of a tall building and the summit of a mountain—size of drops of rain—velocity of their fall—quantity of rain in different latitudes—extraordinary falls of rain—remarks on the rain of this country—influence of the moon—absence of rain—remarkable drought in south america—its terrible effects and consequences—artificial rains.

The quantity of rain which falls at different parts of the earth's surface is very variable; and p. 118for the purpose of measuring it instruments called *Rain-gauges* have been contrived. The simplest form is a funnel three or four inches high, and having an area of one hundred square inches. This may be placed in the mouth of a large bottle, and, after each fall of rain, the quantity may be measured by a glass jar divided into inches and parts. This simple gauge being placed on the ground in an open spot, will evidently represent a portion of the ground, and will show the depth of rain which would cover it at and about that spot, supposing the ground to be horizontal, and that the water could neither flow off nor sink into the soil. Thus, by taking notice of the quantity of rain which falls day by day, and year by year, and taking the average of many years, we get the mean annual quantity of rain for the particular spot in question. By an extension of these observations, it is evident that the mean annual fall of rain may be known for a district or a kingdom.

A more convenient form of rain-gauge than the one just noticed, is made by placing the funnel at the top of a brass or copper cylinder, connected with which at the lower point, is a glass tube with a scale, measuring inches and tenths of an p. 119inch. The water stands at the same height in the glass tube as it does in the cylinder, and being visible in the tube the height can be immediately read on the scale. The cylinder and the tube are so constructed, that the sum of the areas of their sections is a given part, such as a tenth of the area of the mouth of the funnel; so that each inch of water in the tube is equal to the tenth of an inch of water which enters the mouth of the funnel. A stop-cock is added for drawing off the water from the cylinder after each observation is noted down.

Some rain-gauges are constructed for showing the quantity of rain which falls from each of the four principal quarters. Others are made so as to register, themselves, the quantity of rain fallen. One of this kind, by Mr. Crosley, consists of a funnel through which the rain passes to a vibrating trough; when, after a sufficient quantity has fallen into its higher side, it sinks down and discharges the rain which escapes by a tube. The vibrating action of this trough moves a train of wheel-work and indices, which register upon a dial plate the quantity of rain fallen.

Whatever form of rain-gauge is adopted, it p. 120must be placed in an exposed situation, at a distance from all buildings, and trees, and other objects likely to interfere with the free descent of rain into the funnel. It is usual, in rainy weather, to observe the quantity of water in the gauge every morning; but this does not seem to be often enough, considering how freely water evaporates in an exposed situation. An error may also arise from some of the water adhering to the sides of the vessel, unless an allowance is made for the quantity thus lost by a contrivance such as the following: — Let a sponge be made damp, yet so that no water can be squeezed from it, and with this collect all the water which adheres to the funnel and cylinder, after as much as possible has been drawn off; then, if the sponge be squeezed, and the water from it be received in a vessel which admits of measuring its quantity, an estimate may be made of the depth due to it; and this being added to the depth given by the instrument, would probably show correctly the required depth of rain.

When snow has fallen the rain-gauge may not give a correct quantity, as a portion of it may be blown out, or a greater quantity may have fallen p. 121than the mouth will contain. In such cases, it is recommended to take a cylindrical tube and press it perpendicularly into the snow, and it will bring out with it a cylinder equal to the depth. This, when melted, will give the quantity of water which can be measured as before. The proportion of snow to water is about seventeen to one; and hail to water, about eight to one. These quantities, however, may vary according to the circumstances under which the snow or hail has fallen, and the time they have been upon the ground.

The rain-gauge should be placed as near the surface of the ground as possible; for it is a perplexing circumstance, that the rain-gauge indicates very different quantities of rain as falling upon the very same spot, according to the different heights at which it is placed. Thus it has been found, that the annual depth of rain at the top of Westminster Abbey was 12.1 inches nearly, while, on the top of a house sixteen feet lower, it was rather more than 18.1 inches, and on the ground, in the garden of the house, it was 22.6 inches. M. Arago has also found from observations made during twelve years, that on the terrace of the Observatory at Paris the annual depth was about p. 1222< inches less than in the court thirty yards below.

It would naturally be expected from these observations, that less rain falls on high ground than at the level of the sea. Such however is not the case, except on abrupt elevations; where the elevation is made by the natural and gradual slope of the earth's surface, the quantity of rain is greater on the mountain than in the plain. Thus, on the coast of Lancashire, there is an annual fall of 39 inches; while at Easthwaite, among the mountains in the same county, the annual depth of rain amounts to 86 inches. By comparing the registers at Geneva and the convent of the Great St. Bernard, it appears that at the former place, by a mean of thirty-two years, the annual fall of rain is about 30> inches; while at the latter, by a mean of twelve years, it is a little over 60 inches.

In order to explain these remarkable differences, it must not be supposed that the clouds extend down to the ground, so as to cause more rain at the foot of Westminster Abbey than on its roof. There is no doubt that in moist weather the air contains more water near the ground than a few hundred feet above it; and probably, the same cause p. 123which determined a fall from the cloud, would also throw down the moisture floating at a low elevation. Much rain also proceeds from drifting showers, of short duration, and the current moves more slowly along the surface, and allows the drops to fall as fast as they are formed. In hilly countries, on the contrary, clouds and vapours rest on the summits without descending into the plains, and, according to some, the hills attract electricity from the clouds, and thus occasion rain to fall. Mr. Phillips supposes that each drop of rain continues to increase in size from the commencement to the end of its descent, and as it passes successively through the moist strata of the air, obtains its increase from them; while the rain which falls on the mountain may leave these moist strata untouched, so that they may, in fact, not form rain at all.

The drops of rain are of unequal size, as may be seen from the marks made by the first drops of a shower upon any smooth surface. They vary in size from perhaps the twenty-fifth to a quarter of an inch in diameter. It is supposed that in parting from the clouds they fall with increasing speed, until the increasing resistance of the air p. 124becomes equal to their weight, when they continue to fall with an uniform velocity. A thunder-shower pours down much faster than a drizzling rain. A flake of snow, being perhaps nine times more expanded than water, descends thrice as slow. But hailstones are often several inches in length, and fall with a velocity of seventy feet in a second, or at the rate of about fifty miles an hour, and hence the destructive power of these missiles in stripping and tearing off fruit and foliage.

The annual quantity of rain decreases from the equator to the poles, as appears from the following table, which gives the name of the station, its latitude, and the average annual number of inches of rain:—

Coast of Malabar	lat. 110 30′ N.	135= inches.
At Grenada, Antilles	120	126
At Cape Frangois, St. Domingo	190 46′	120
At Calcutta	220 23′	81
At Rome	410 54′	39
In England	50 to 550	31

At St. Petersburgh	59° 16'	16
At Uleaborg	65° 30'	13=

The number of rainy days, on the contrary, increases from the equator to the poles.

p. 125 From 12° to 43° N. lat.—the number of rainy days in the year amounts to	78
From 43° to 46°	103
From 46° to 50°	134
From 50° to 60°	161

The greatest depth of rain which falls in the Indian ocean is during the time when the periodical winds, called the *monsoons*, change their direction. When the winds blow directly in-shore the rains are very abundant, so much so that, after a continuance of twenty-four hours, the surface of the sea has been covered with a stratum of fresh water, good enough for drinking, and ships have actually filled their casks from it. Colonel Sykes observes, that the deluge-like character of a monsoon in the Gh`ts of Western India, is attested by the annual amount of 302< inches, at Malcolmpait, on the Mahabuleshwar Hills.

A great depth of rain in a short time has occasionally been witnessed in Europe. At Genoa, on the 25th of October, 1822, a depth of thirty inches of rain fell in one day. At Joyeuse, on the 9th of October, 1827, thirty-one inches of rain fell in twenty-two hours. Previous to the great floods of Moray, in 1829, the rain is described as p. 126 being so thick that the very air itself seemed to be descending in one mass of water upon the earth. Nothing could withstand it. The best finished windows were ineffectual against it, and every room exposed to the north-east was deluged. The smaller animals, the birds, and especially game, of all kinds, were destroyed in great numbers by the rain alone, and the mother partridge, with her brood and her mate, were found chilled to death amidst the drenching wet. It was also noticed, that, as soon as the flood touched the foundation of a dry stone wall, the sods on the top of it became as it were alive with mice, all forcing their way out to escape from the inundation which threatened their citadel; and in the stables, where the water was three feet deep, rats and moles were swimming about among the buildings.

Among the Andes it is said to rain perpetually; but in Peru it never rains, moisture being supplied during a part of the year by thick fogs, called *garuas*. In Egypt, and some parts of Arabia, it seldom rains at all, but the dews are heavy, and supply with moisture the few plants of the sandy regions.

There is a great variation in the quantity of p. 127rain that falls in the same latitude, on the different sides of the same continent, and particularly of the same island. The mean fall of rain at Edinburgh, on the eastern coast, is 26 inches; while at Glasgow, on the western coast, in nearly the same latitude, it is 40 inches. At North Shields, on the eastern coast, it is 25 inches; while at Coniston, in Lancashire, in nearly the same latitude, on the western coast, it is 85 inches.

The amount of rain in a district may be changed by destroying or forming forests, and by the inclosure and drainage of land. By thinning off the wood in the neighbourhood of Marseilles, there has been a striking decrease of rain in fifty years.

In Mr. Howard's observations on the climate of this country, he has found, on an average of years, that it rains every other day; that more rain falls in the night than in the day; that the greatest quantity of rain falls in autumn, and the least in winter; that the quantity which falls in autumn is nearly double that in spring; that most rain falls in October and least in February, and that May comes nearest to the mean: that one year in every five, in this country, may be expected to be extremely dry, and one in ten extremely wet.

p. 128According to Dalton, the mean annual amount of rain and dew for England and Wales is 36 inches. The mean all over the globe is stated to be 34 inches.

There seems to be some real connexion between the changes of the moon and the weather. Mr. Daniell says, "No observation is more general; and on no occasion, perhaps, is the almanac so frequently consulted as in forming conjectures upon the state of the weather. The common remark, however, goes no further than that changes from wet to dry, and from dry to wet, generally happen at the changes of the moon. When to this result of universal experience we add the philosophical reasons for the existence of tides in the akrial ocean, we cannot doubt that such a connexion exists. The subject, however, is involved in much obscurity." At Viviers, it was observed that the number of rainy days was greatest at the first quarter, and least at the last. Mr. Howard has observed that, in this country, when the moon has south declination, there falls but a moderate quantity of rain, and that the quantity increases till she has attained the greatest northern declination. He thinks there is "evidence of a great *tidal wave*, p. 129or swell in the atmosphere, caused by the moon's attraction, preceding her in her approach to us, and following slowly as she departs from these latitudes."

Most dry climates are subject to periodical droughts. In Australia, they return after every ten or twelve years, and are then followed by excessive rains, which gradually become less and less till another drought is the consequence.

When Mr. Darwin was in South America, he passed through a district which had long been suffering from dry weather. The first rain that had fallen during that year was on the 17th of May, when it rained lightly for about five hours. "With this shower," he says, "the farmers, who plant corn near the sea-coast, where the atmosphere is more humid, would break up the ground; with a second, put the seed in; and, if a third should fall, they would reap in the spring a good harvest. It was interesting to watch the effect of this trifling amount of moisture. Twelve hours afterwards the ground appeared as dry as ever; yet, after an interval of ten days, all the hills were faintly tinged with green patches; the grass being sparingly scattered in hair-like fibres a full inch in p. 130length. Before this shower every part of the surface was bare as on a high road."

A fortnight after this shower had fallen, Mr. Darwin took an excursion to a part of the country to which the shower had not extended. "We had, therefore," he says, "in the first part of our journey a most faint tinge of green, which soon faded away. Even where brightest, it was scarcely sufficient to remind one of the fresh turf and budding flowers during the spring of other countries. While travelling through these deserts, one feels like a prisoner, shut up in a gloomy courtyard, longing to see something green, and to smell a moist atmosphere."

The effects of a great drought in the Pampas are thus described. "The period included between the years 1827 and 1830 is called the 'gran seco' or the great drought. During this time so little rain fell, that the vegetation, even to the thistles, failed; the brooks were dried up, and the whole country assumed the appearance of a dusty high road. This was especially the case in the northern part of the province of Buenos Ayres, and the southern part of St. Fe. Very great numbers of birds, wild animals, cattle, and horses, p. 131perished from the want of food and water. A man told me that the deer used to come into his courtyard to the well which he had been obliged to dig to supply his own family with water; and that the partridges had hardly strength to fly away when pursued. The lowest estimation of the loss of cattle in the province of Buenos Ayres alone, was taken at one million head. A proprietor at San Pedro had previously to these years 20,000 cattle; at the end not one remained. San Pedro is situated in the midst of the finest country, and even now again abounds with animals; yet, during the latter part of the 'gran seco' live cattle were brought in vessels for the consumption of the inhabitants. The animals roamed from their *estancias*, and wandering far to the southward, were mingled together in such multitudes that a government commission was sent from Buenos Ayres to settle the disputes of the owners. Sir Woodbine Parish informed me of another and very curious source of dispute; the ground being so long dry, such quantities of dust were blown about, that in this open country the landmarks became obliterated, and people could not tell the limits of their estates.

p. 132"I was informed by an eye-witness, that the cattle in herds of thousands rushed into the river Parana, and being exhausted by hunger they were unable to crawl up the muddy banks, and thus were drowned. The arm which runs by San Pedro was so full of putrid carcasses, that the master of a vessel told me, that the smell rendered it quite impossible to pass that way. Without doubt, several hundred thousand animals thus perished in the river. Their bodies, when putrid, floated down the stream, and many in all probability were deposited in the estuary of the Plata. All the small rivers became highly saline, and this caused the death of vast numbers in particular spots, for when an animal drinks of such water it does not recover. I noticed, but probably it was the effect of a gradual increase, rather than of any one period, that the smaller streams in the Pampas were paved with bones. Subsequently to this unusual drought, a very rainy season commenced, which caused great floods. Hence it is almost certain, that some thousands of these skeletons were buried by the deposits of the very next year. What would be the opinion of a geologist viewing such an enormous collection of bones, of all kinds p. 133of animals and of all ages, thus embedded in one thick earthy mass? Would he not attribute it to a flood having crept over the surface of the land, rather than to the common order of things?"

Captain Owen mentions a curious effect of a drought on the elephants at Benguela on the western coast of Africa:—"A number of these animals had some time since entered the town in a body to possess themselves of the wells, not being able to procure any water in the country. The inhabitants mustered, when a desperate conflict ensued, which terminated in the ultimate discomfiture of the invaders, but not until they had killed one man, and wounded several others." The town is said to have a population of nearly three thousand. Dr. Malcolmson states, that during a great drought in India the wild animals entered the tents of some troops at Ellore, and that a hare drank out of a vessel held by the adjutant of the regiment.

In connexion with droughts may be mentioned a plan [133] proposed by Mr. Espy of the United States of America, for remedying them by means p. 134of *artificial rains*. That gentleman says, that if a large body of heated air be made to ascend in a column, a large cloud will be generated, and that such cloud will contain in itself a self-sustaining power, which may move from the place over which it was formed, and cause the air over which it passes to rise up into it and thus form more cloud and rain, until the rain may become general.

It is proposed to form this ascending column of air by kindling large fires which, Mr. Espy says, are known to produce rain. Humboldt speaks of a mysterious connexion between volcanoes and rain, and says that when a volcano bursts out in South America in a dry season, it sometimes changes it to a rainy one. The Indians of Paraguay, when their crops are threatened by drought, set fire to the vast plains with the intention of producing rain. In Louisiana, heavy rains have been known from time immemorial to succeed the conflagration of the prairies; and the inhabitants of Nova Scotia bear testimony to a similar result from the burning of their forests. Great battles are said to produce rain, and it is even stated that the spread of manufactures in a particular district deteriorates the climate of such district, the ascending p. 135current occasioned by the tall chimney of every manufactory tending to produce rain. In Manchester, for example, it is said to rain six days out of seven.

p. 136

p. 137CHAPTER VI.

the rainbow—decomposition of white light by the prism—formation of primary and secondary bows—rainbows in mountain regions—the rainbow a sacred emblem—lunar rainbow—light decomposed by clouds—their beautiful colours—examples.

By means of rain and rain clouds we get that beautiful appearance so well known as the rainbow. In order to form some idea of the manner in which the rainbow is produced, it is necessary to know something of the manner in which light is composed. Sir Isaac Newton was the first philosopher who clearly explained the composition of light, as derived from the sun. He admitted a ray of the sun into a darkened room through a small hole in the window shutters; in front of this hole he placed a glass prism, and at a considerable distance behind the prism he placed a white screen. If there had been no prism between the hole and the screen, the ray of light would have proceeded in the direction of the dotted lines, and p. 138a bright spot would have fallen upon the floor of the room, as shown in the figure. But the effect of the prism is to refract or bend the ray out of its ordinary course, and in doing so it does not produce a white spot upon the screen, but a long streak of beautiful colours, in the order marked in the figure, red being at the bottom, then orange, yellow, green, blue, indigo, and violet at the top.

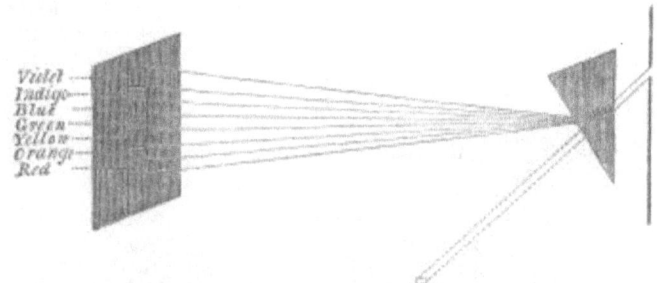

In order to account for the production of these colours from a ray of light, Newton supposed that such a ray is actually made up of seven distinct colours, which being mixed in proper proportions neutralize or destroy each other. In order to account for the decomposition of the ray of white light by the prism, and for the lengthened form of the *spectrum*, as it is called, he supposed that each of the seven coloured rays was capable of being bent by the prism in a different manner from the rest. Thus, in the figure, the red appears to be less bent out of the direction of the original ray than the orange—the orange less than the yellow, and so on until we arrive at the violet, which is bent most of all.

It is scarcely necessary to remark, that these views were found to be correct, except as regards the number of colours in the solar spectrum; for it is now ascertained, with tolerable certainty, that there are only three primitive or pure colours in nature, and these are *red*, *yellow*, and *blue*; and it is supposed that by mingling two or more of these colours in various proportions, all the colours in nature are produced.

Now, to apply this explanation to the production of the rainbow, which is usually seen under the following circumstances:—The observer is placed with his back to the sun, and at some distance before him rain is falling,—the air between the sun and the rain being tolerably clear. He then often sees two circular arcs or bows immediately in front of him. The colours of the inner bow are the more striking and vivid of the two. Each exhibits the same series of colours as in the spectrum formed by the prism; namely, *red*, *orange*, *yellow*, *green*, *blue*, *indigo*, and *violet*; but the arrangement of these colours is different in the two bows, for while in the inner bow the lower edge is violet and the upper red, in the outer bow the lower edge is red and the upper violet. The production of both bows is due to the refraction and reflexion of light, the drops of rain forming, in fact, the prism which decomposes the white light of the sun. The colours in the rainbow have the same proportional breadth as the spaces in the prismatic spectrum. "The bow is, therefore," as Sir D. Brewster remarks, "only an infinite number of prismatic spectra, arranged in the circumference of a circle; and it would be easy, by a circular arrangement of prisms, or by covering up all the central part of a large lens, to produce a small arch of exactly the same colours. All we require, therefore, to form a rainbow, is a great number of transparent bodies capable of forming a great number of prismatic spectra from the light of the sun."

The manner in which the drops of rain act as prisms, may, perhaps, be better understood with p. 141the assistance of the following diagram. Suppose the two lower circles to represent drops of rain which assist in forming the primary bow, and the two upper circles similar drops which help to produce the secondary bow; and let S represent rays of the sun falling upon them. The rays of the sun fall upon every part of the drop; but, as those p. 142which pass through or near the centre come out on the opposite side and form a focus, they need not be taken into account. Those rays, however, which fall on the upper side of the drops, will be bent or refracted, the red rays least, and the violet most; and will fall upon the back of the drop in such a manner as to be reflected to the under part of the drop; on quitting which they will be again refracted, so as to be seen at E, where there will appear to the observer a prismatic spectrum with the red uppermost, and the violet undermost. These remarks apply to those drops only which form the upper part of the bow, but it is obvious that a similar reasoning applied to the drops to the right and left of the observer, will complete the p. 143bow. The inclination of the red ray and the violet ray to the sun's rays, is 420 2' for the red, and 400 17' for the violet, so that the breadth of the primary bow is 10 45'.

Thus it will be seen, that the primary bow is produced by two refractions, and one intermediate reflection of the rays that fall on the upper sides of the drops of rain. It is different with the rays which enter the drops below. The red and violet rays will be bent or refracted in different directions; and, after being twice reflected, will be again bent towards the eye of the observer at E; but in this case the violet forms the upper part, and the red the under part of the spectrum. The inclination of these rays to the sun's rays at S, is 500 58' for the red ray, and 540 10' for the violet ray; so that the breadth of the bow is 30 10', and the distance between the primary and secondary bows is p. 14480 15'. Hence the secondary is formed in the outside of the primary bow, with its colours reversed, in consequence of their being produced by two reflexions and two refractions. The colours of the secondary bow are much fainter than those of the primary, because they undergo two reflexions instead of one.

There is something very wonderful in the rapidity and perfection with which these natural prisms, the falling drops of rain, produce these effects. In the inconceivably short space of time occupied by a drop falling through those parts of the sky which form the proper angles with the sun's rays and the eye of the observer, the light enters the surface of the drop, undergoes within it one or two reflexions, two refractions and decompositions, and has reached the eye; and all this is done in a portion of time too small for the drop to have fallen through a space which we have the means of measuring.

It will be understood, that since the eyes of different observers cannot be in precisely the same place at the same time, no two observers can see the *same* rainbow; that is to say, the bow produced by one set of drops to the eye of one observer is p. 145produced by another set of drops to the eye of another observer.

A rainbow can never be greater than a semicircle, unless the spectator is on elevated ground; for if it were greater than a semicircle the centre of the bow would be above the horizon, while the sun, which must be in a line drawn through that centre and the eye of the observer, would be below the horizon: but in such a case, the sun could not shine on the drops of rain, and consequently there could be no rainbow.

When the rain cloud is of small extent only a portion of a bow is visible; when the cloud overspreads a large part of the sky a perfect bow appears. Sometimes the bow may be traced across a portion of blue sky, or it may appear to rest on the ground. In the former case, there are vapours in the air too thin to be seen, but sufficient to refract and reflect the rays of light; in the latter, the drops of rain, adhering to the grass and foliage, produce the same effect. A coloured bow, similar to that produced by rain, is sometimes seen in the spray of a fountain or of a water-fall, and also in mists that lie low upon the ground.

In mountainous and stormy regions rainbows p. 146are often seen to great advantage. In the islands off the Irish coast the author of "Letters from the Irish Islands," describes the rainbow of winter "as gradually advancing before the lowering clouds, sweeping with majestic stride across the troubled ocean, then, as it gained the beach, and seemed almost within one's grasp, vanishing amid the storm of which it had been the lovely but treacherous forerunner. It is, I suppose, a consequence of our situation, and the close connexion between sea and mountain, that the rainbows here are so frequent and so peculiarly beautiful. Of an amazing breadth, and of colours vivid beyond description, I know not whether most to admire this akrial phenomenon, when suspended in the western sky, one end of the bow sinks behind the Island of Boffin, while at the distance of several leagues the other rests upon the misty hills of Ennis Turc; or when, at a later hour of the day, it has appeared stretched across the ample sides of Mulbrea, penetrating far into the deep blue waters that flow at its base. With feelings of grateful recollection, too, we may hail the repeated visits of this heavenly messenger, occasionally as often as five or six times in the course of the same p. 147day, in a country exposed to such astonishing, and, at times, almost incessant floods of rain."

The beauty of the rainbow is not the only reason why we should regard it with interest. The rainbow was appointed by God himself as a sign of the covenant of mercy, made with Noah and with all mankind, after the flood. The words in which this declaration was made to mankind, are recorded in the Book of Genesis, chap. ix. ver. 11 to 16.

Burnet, in his "Sacred Theory of the Earth," has some remarks on the first appearance of the rainbow to the inhabitants of the earth after the deluge. He says, "How proper and how apposite a sign would this be for Providence to pitch upon, to confirm the promise made to Noah and his posterity, that the world should be no more destroyed by water! It had a secret connexion with the effect itself, and was so far a natural sign; but, however, appearing first after the deluge, and in a watery cloud, there was, methinks, a great easiness and propriety of application for such a purpose. And if we suppose, that while God Almighty was declaring his promise to Noah, and the sign of it, there appeared at the same time in the clouds p. 148a fair rainbow, that marvellous and beautiful meteor which Noah had never seen before; it could not but make a most lively impression upon him, quickening his faith, and giving him comfort and assurance that God would be stedfast to his promise."

A rainbow is sometimes formed by the rays of the moon falling upon drops of rain, in the same manner as the solar rays, and refracted and reflected by the drops; but the colours are faint in consequence of the feeble light of the moon compared with that of the sun. A lunar rainbow has been thus described by an observer:—"The moon was truly 'walking in brightness,' brilliant as she could be, not a cloud was to be seen near her; and over against her, toward the north-west, or perhaps rather more to the north, was a rainbow, a vast arch, perfect in all its parts, not interrupted or broken as rainbows frequently are, but unremittedly visible from one horizon to the other. In order to give some idea of its extent, it is necessary to say, that, as I stood toward the western extremity of the parish of Stoke Newington, it seemed to take its rise from the west of Hampstead, and to end perhaps in the river Lea, the p. 149eastern boundary of Tottenham. Its colour was white, cloudy, or greyish, but a part of its western limb seemed to exhibit tints of a faint sickly green. After some time the moon became darkened by clouds, and the rainbow of course vanished."

The brilliant colours of the solar rainbow are frequently produced by the clouds without any prismatic arrangement. The light of the sun is decomposed by a process called absorption: for example, white light is composed of red, yellow, and blue rays, in certain proportions; now, if in passing through, or falling upon any substance whatever, the red rays are stifled or absorbed, while the yellow and blue are allowed to pass or to be reflected, it is obvious that such a substance cannot appear white, because one of the elements of white light, namely, the red, is wanting; it must therefore appear of such a colour as results from the combination of yellow and blue; the substance will therefore appear green. So, also, when white light falls upon what we call a *red* surface, the yellow and blue rays are stifled or absorbed, leaving the red only to be reflected. Now, when we consider the various ways in which this absorption may take place; one or two, or all of the coloured rays being absorbed in every possible proportion, it is easy to form some idea of the manner by which the innumerable tints of the sky are produced.

It has been calculated, that, of the horizontal sunbeams which pass through two hundred miles of air, scarcely a two thousandth part reaches the earth. A densely formed cloud must therefore detain a much larger share; and those dark and sombre forms, which sometimes make the sky so gloomy, can only result from the abundant absorption of the solar light. The brilliant whiteness which their edges occasionally exhibit, must result from the more copious transmission of light, so that the depths of shade in a cloud may be regarded as comparative measures of the varied thickness of its mass.

Sometimes the clouds absorb equally all the solar rays, in which case the sun and moon appear through them perfectly white. Instances are recorded in which the sun appeared of a pale blue. It has also been observed to be orange at its upper part, while the lower was of a brilliant red.

The position from which clouds are seen, has much to do with their colours; and it seems difficult sometimes to believe that the clouds, which in the evening are seen drenched with crimson and gold, are the same we beheld absolutely colourless in the middle of the day.

In the immediate neighbourhood of the sun the p. 152most brilliant colours may be disclosed; and their vividness and intensity diminish, and at last disappear at some distance from it. Parry noticed some white fleecy clouds, which, at the distance of fifteen or twenty degrees from the sun, reflected from their edges the most soft and tender tints of yellow, bluish green, and lake; and as the clouds advanced the colours increased gradually, until they reached a sort of limit two degrees below the solar orb. As the current continued to transport them, the vividness of colour became weakened by almost insensible degrees until the whole assemblage of tints vanished.

"Who can venture to imitate, by the pencil, the endless varieties of red and orange and yellow which the setting sun discloses, and the magical illusions which all the day diversify the vast and varied space the eye travels over in rising gradually from the horizon to the upper sky? Those who have paid any attention to colours, must be aware of the difficulty of describing the various tints and shades that appear, and which are known to amount to many thousands."

The rapid changes of colour which the clouds p. 153undergo, seem to depend on something more than change of position either in the cloud or in the sun. Forster mentions an instance of some detached cirro-cumuli being of a fine golden yellow, but in a single minute becoming deep red. On another occasion he saw the exact counterpart in a cirro-stratus, by its instantly changing from a beautiful red to a bright golden yellow. "What, indeed, can be more interesting, than when by the breaking out of the sun in gleams, a cloud which a moment before seemed only an unshapened mass devoid of all interest and beauty, is suddenly pierced by cataracts of light, and imbued with the most splendid colours, varying every instant in intensity? Numerous examples occur of this beautiful play of colour, which cannot but remind us of the phenomena displayed by the pigeon's neck and the peacock's tail, by opal and pearl.

"After the sun is set, the mild glow of his rays is still diffused over every part; and it has been remarked, that the clouds assume their brightest and most splendid colours a few minutes after it is below the horizon. It is in the finest weather that p. 154the colouring of the sky presents the most perfect examples of harmony, in tempestuous weather it being almost always inharmonious. At the time of a warm sun-setting, the whole hemisphere is influenced by the prevailing colour of the light. The snowy summits of the Alps appear about sunset of a most beautiful violet colour, approaching to light crimson or pink. It is remarkable, also, as an example of that general harmony which prevails in the material world, that the most glowing and magnificent skies occur when terrestrial objects put on their deepest and most splendid hues. It has also been observed, that it is not the change of vegetation only, which gives to the decaying charms of autumn their finest and most golden hues, but also the atmosphere and the peculiar lights and shadows which then prevail; and there can be no doubt, on the other hand, that our perception of beauty in the sky is very much influenced by the surrounding scenery. In autumn all is matured; and the rich hues of the ripened fruits and the changing foliage are rendered still more lovely by the warm haze which a fine day at that season presents. So, also, the earlier hues of p. 155spring have a transparency, and a thousand quivering lights, which in their turn harmonize with the light and flitting clouds and uncertain shadows which then prevail." [155]

p. 156

CHAPTER VII.

remarkable showers—showers of sand—of mud—showers of sulphur, or yellow rain—luminous rain—red rain, or showers of blood—superstitions connected therewith—explanation of the cause—showers of fish—showers of rats—showers of frogs—insect shower—showers of vegetable substances—manna—wheat—showers of stones—meteoric stones, or aerolites—meteoric iron—suppositions respecting them—fossil rain.

Water, in the state of rain, hail, snow, or dew, is generally the only substance which falls from the atmosphere upon the earth. There are, however, many well authenticated instances of various substances being showered down upon the land, to the great alarm of persons who were ignorant that the powerful action of the wind was, perhaps, the chief cause of the strange visitations to which we allude.

We read of showers of sand, mud, sulphur, blood, fishes, frogs, insects, and stones; and it may be useful, as well as interesting, to quote a few examples of each description of shower.

On the west coast of Africa, between Cape Bojador and Cape Verd, and thence outwards, the land, during the dry season, consists of little else but dust or sand, which, on account of its extreme fineness, is raised into the atmosphere by the slightest current of air; while a moderate wind will convey it to so considerable a distance as even to annoy ships crossing the Atlantic. On the 14th and 15th January, 1839, the Prussian ship, *Princess Louisa*, being in N. lat. 240 20', and W. long. 260 42', had her sails made quite yellow by the fine sand which covered them. This effect was produced when the distance from land was as much as from 120 to 200. About a fortnight after the time when this ship crossed these parts of the Atlantic, a similar effect was produced on board the English ship *Roxburgh*. One of the passengers, the Rev. W. B. Clarke, says:—"The sky was overcast, and the weather thick and insufferably oppressive, though the thermometer was only 720. At 3 p.m. Feb. 4, the wind suddenly lulled into a calm; then rose from the SW. accompanied by rain, and the air appeared to be filled with dust, which affected the eyes of the passengers and crew. The weather was clear and fine, and the powder which covered the sails was of a reddish-brown colour, resembling the ashes ejected from Vesuvius; and Mr. Clarke thinks that this dust may have proceeded from the volcanic island of Fogo, one of the Cape de Verds, about forty-five miles from the place where the ship then was.

In countries which are subject to long-continued droughts the soil is frequently converted into dust, which, being carried away by the winds, leaves the land barren. The climate of Buenos Ayres, in South America, has of late years been subject to such droughts, as to disappoint the hopes of the husbandman and the breeder of cattle. In the early part of 1832, the drought had reached to such a height as to convert the whole province into one continued bleak and dreary desert. The clouds of dust raised by the winds were so dense as completely to obscure the sun at mid-day, and envelope the inhabitants in almost total darkness. When the rains at length commenced, in March, the water, in its passage through the air, intermingled so completely with the dust suspended in it, as to descend in the form of showers of mud; and, p. 160on some occasions, gave to the whole exterior of the houses the appearance of having been plastered over with earth. Many flocks of sheep were smothered on these occasions, in a similar manner as in the snow-storms which occur in the mountainous districts of Scotland.

Showers of sulphur, or yellow rain, have fallen at different times in various parts of Europe; and sometimes, when falling by night, they have appeared luminous, to the great alarm of the observers. Yellow rain has been accounted for in the following way:—The pollen, or impregnating seed-dust of the flowers of the fir, birch, juniper, and other trees, is of a yellow colour, and this pollen, by the action of the wind, is carried to a considerable distance, and descends with falling rain. This yellow rain has also been found impregnated with sulphur; and during a shower of this kind which once fell in Germany, matches were made by being dipped in it.

Many examples of luminous rain are recorded on good authority. One of the latest instances is mentioned by Dr. Morel Deville, of Paris, who on the 1st of November, 1844, at half-past eight o'clock in the evening, during a heavy fall of rain, p. 161noticed, as he was crossing the court of the College Louis-le-Grand, that the drops, on coming in contact with the ground, emitted sparks and tufts (*aigrettes*) of light, accompanied by a rustling and crackling noise; a smell of phosphorus having been immediately after perceptible. The phenomenon was seen three times. At the same hour a remarkable brightness was seen in the northern sky.

An officer of the Algerian army states, that during a violent storm on the 20th September, 1840, the drops of rain that fell on the beards and mustachios of the men were luminous. When the hair was wiped the appearance ceased; but was renewed the moment any fresh drops fell on it.

But of all these remarkable showers, the greatest alarm has been occasioned by *red rain*, or showers of blood as they have been ignorantly called. In the year 1608, considerable alarm was excited in the city of Aix and its vicinity by the appearance of large red drops upon the walls of the cemetery of the greater church, which is near the walls of the city, upon the walls of the city itself, and also upon the walls of villas, hamlets, and towns, for some miles round the city. The husbandmen are p. 162said to have been so alarmed, that they left their labour in the fields and fled for safety into the neighbouring houses; and a report was set on foot, that the appearance was produced by demons or witches shedding the blood of innocent babes. M. Peiresc, thinking this story of a bloody shower to be scarcely reconcileable with the goodness and providence of God, accidentally discovered, as he thought, the true cause of the phenomenon. He had found, some months before, a chrysalis of remarkable size and form, which he had enclosed in a box; he thought no more of it, until hearing a buzz within the box, he opened it, and perceived that the chrysalis had been changed into a beautiful butterfly, which immediately flew away, leaving at the bottom of the box a red drop of the size of a shilling. As this happened about the time when the shower was supposed to have fallen, and when multitudes of those insects were observed fluttering through the air in every direction, he concluded that the drops in question were emitted by them when they alighted upon the walls. He, therefore, examined the drops again, and remarked that they were not upon the upper surfaces of stones and buildings, p. 163as they would have been if a shower of blood had fallen from the sky, but rather in cavities and holes where insects might nestle. He also noticed that they were to be seen upon the walls of those houses only which were near the fields; and not upon the more elevated parts of them, but only up to the same moderate height at which butterflies were accustomed to flutter. This was, no doubt, the correct explanation of the phenomenon in question; for it is a curious and well-ascertained fact, that when insects are evolved from the pupa state, they always discharge some substance, which, in many butterflies, is of a red colour, resembling blood, while in several moths it is orange or whitish.

It appears, however, from the researches of M. Ehrenberg, a distinguished microscopic observer, that the appearances of blood which have at different times been observed in Arabia, Siberia, and other places, are not to be attributed to one, but to various causes. From his account, it appears that rivers have flowed suddenly with red or bloody water, without any previous rain of that colour having fallen; that lakes or stagnant-waters were suddenly or gradually coloured without p. 164previous blood-rain; that dew, rain, snow, hail, and shot-stars, occasionally fall from the air red-coloured, as blood-dew, blood-rain, and clotted blood; and, lastly, that the atmosphere is occasionally loaded with red dust, by which the rain accidentally assumes the appearance of blood-rain, in consequence of which rivers and stagnant waters assume a red colour.

The blood-red colour sometimes exhibited by pools, was first satisfactorily explained at the close of the last century. Girod Chantran, observing the water of a pond to be of a brilliant red colour, examined it with the microscope, and found that the sanguine hue resulted from the presence of innumerable animalculf, not visible to the naked eye. But, before this investigation, Linnfus and other naturalists had shown that red infusoria were capable of giving that colour to water which, in early times, and still, we fear, in remote districts, was supposed to forebode great calamities. In the year 1815 an instance of this superstitious dread occurred in the south of Prussia. A number of red, violet, or grass-green spots were observed in a lake near Lubotin, about the end of harvest. In winter the ice was coloured in the same manner at the surface, p. 165while beneath it was colourless. The inhabitants, in great dismay, anticipated a variety of disasters from the appearance; but it fortunately happened that the celebrated chemist Klaproth, hearing of the circumstance, undertook an examination of the waters of the lake. He found them to contain an albuminous vegetable matter, with a particular colouring matter similar to indigo, produced, probably, by the decomposition of vegetables in harvest; while the change of colour from green to violet and red, he explained by the absorption of more or less oxygen. A few years ago the blood-red waters of a Siberian lake were carefully examined by M. Ehrenberg, and found to contain multitudes of infusoria, by the presence of which this remarkable appearance was accounted for. Thus it appears that both animals and vegetables are concerned in giving a peculiar tint to water. It has also been ascertained that red snow is chiefly occasioned by the presence of red animalculf.

Showers of fish and frogs are by no means uncommon, especially in India. One of these showers, which fell about twenty miles south of Calcutta, is thus noticed by an observer:—"About two o'clock, p.m., of the 20th inst., (Sept. 1839,) we had a very smart shower of rain, and with it descended a quantity of live fish, about three inches in length, and all of one kind only. They fell in a straight line on the road from my house to the tank which is about forty or fifty yards distant. Those which fell on the hard ground were, as a matter of course, killed from the fall, but those which fell where there was grass sustained no injury; and I picked up a large quantity of them, 'alive and kicking,' and let them go into my tank. The most strange thing that ever struck me in connexion with this event, was, that the fish did not fall helter skelter, everywhere, or 'here and there;' but they fell in a straight line, not more than a cubit in breadth." Another shower is said to have taken place at a village near Allahabad, in the month of May. About noon, the wind being in the west, and a few distant clouds visible, a blast of high wind came on, accompanied with so much dust as to change the tint of the atmosphere to a reddish hue. The blast appeared to extend in breadth four hundred yards, and was so violent that many large trees were blown down. When the storm had passed over, the ground, south of the village, was found to be covered with fish, not less than three or four thousand in number. They all belonged to a species well known in India, and were about a span in length. They were all dead and dry.

It would be easy to multiply these examples to almost any extent, although they are not so frequent in Great Britain. It is related in Hasted's History of Kent, that about Easter, 1666, in the parish of Stanstead, which is a considerable distance from the sea, and a place where there are no fishponds, and rather a scarcity of water, a pasture field was scattered all over with small fish, supposed to have been rained down during a thunder-storm. Several of these fish were sold publicly at Maidstone and Dartford. In the year 1830, the inhabitants of the island of Ula, in Argyleshire, after a day of very hard rain, which occurred on the 9th March, were surprised to find numbers of small herrings strewed over the fields, perfectly fresh and some of them alive. Some years ago, during a strong gale, herrings and other fish were carried from the Frith of Forth so far as Loch-Leven.

In some countries rats migrate in vast numbers from the high to the low countries; and it is p. 168recorded in the history of Norway, that a shower of these, transported by the wind, fell in an adjacent valley.

Several notices have, from time to time, been brought before the French Academy, of showers of frogs having fallen in different parts of France. Professor Pontus, of Cahors, states, that in August, 1804, while distant three leagues from Toulouse, the sky being clear, suddenly a very thick cloud covered the horizon, and thunder and lightning came on. The cloud burst over the road about sixty toises (383 feet) from the place where M. Pontus was. Two gentlemen, returning from Toulouse, were surprised by being exposed not only to a storm, but to a shower of frogs. Pontus states that he saw the young frogs on their cloaks. When the diligence in which he was travelling, arrived at the place where the storm burst, the road, and the fields alongside of it, were observed full of frogs, in three or four layers placed one above the other. The feet of the horses and the wheels of the carriage killed thousands. The diligence travelled for a quarter of an hour, at least, along this living road, the horses being at a trot.

p. 169In the "Journal de St. Petersburg," is given an account of the fall of a shower of insects during a snow-storm in Russia. "On the 17th October, 1827, there fell in the district of Rjev, in the government of Tver, a heavy shower of snow, in the space of about ten versts (nearly seven English miles), which contained the village of Pakroff and its environs. It was accompanied in its fall by a prodigious quantity of worms of a black colour, ringed, and in length about an inch and a quarter. The head of these insects was flat and shining, furnished with antennf, and the hair in the form of whiskers; while the body, from the head to about one-third of their length, resembled a band of black velvet. They had on each side three feet, by means of which they appeared to crawl very fast upon the snow, and assembled in groups about the plants and the holes in trees and buildings. Several having been exposed to the air in a vessel filled with snow, lived there till the 26th October; although, in that interval, the thermometer had fallen to eight degrees below zero. Some others which had been frozen continued alive equally long; for they were not found exactly encrusted with the ice, but they had formed p. 170round their bodies a space similar to the hollow of a tree. When they were plunged into water they swam about as if they had received no injury; but those which were carried into a warm place perished in a few minutes."

All these remarkable showers may be accounted for, when we consider the mighty power of the wind; especially that form of it which is popularly called the whirlwind. It is now pretty well ascertained, that in all, or most of the great storms which agitate the atmosphere, the wind has a circular or rotatory movement; and the same is probably the case in many of the lesser storms, in which the air is whirled upwards in a spiral curve with great velocity, carrying up any small bodies which may come within the circuit. When such a storm happens at sea, the water-spout is produced. In the deserts of Arabia, pillars of sand are formed; and, in other places various light bodies are caught up; fishponds have been entirely emptied in an instant, and the moving column, whether of water, sand, or air, travels with the wind with great swiftness. When, however, the storm has subsided, the various substances thus caught up and sustained p. 171in the air, are deposited at great distances from the place where they were first found, and thus produce these remarkable showers. In some cases, however, the direct force of the wind has actually blown small fish out of the water, and conveyed them several miles inland.

Showers of nutritious substances have been recorded on good authority. We do not here refer to the manna which fell in such abundance about the Hebrew camp, for that was a miracle specially wrought by the Almighty for the preservation of his chosen people; but, it may be noticed here, that in Arabia, a substance, called "manna," is found in great abundance on the leaves of many trees and herbs, and may be gathered and removed by the wind to a distance. A shower of this kind occurred in 1824. In 1828, a substance was exhibited at the French Academy, which fell in the plains of Persia. It was eaten, and afforded nourishment to cattle, and many other animals; and, on examination, proved to be a vegetable,—the *Lichen esculentus*,—which had been conveyed thither by the winds.

In the Minutes of the proceedings of the Royal Society, 26th June, 1661, we find the following curious narration:—

> p. 172"Col. Tuke brought, in writing, the following *brief account of the supposed rain of wheat*, which was registered:—
>
> "On the 30th of May, 1661, Mr. Henry Puckering, son to Sir Henry Puckering, of Warwick, brought some papers of seeds, resembling wheat, to the king, with a letter written by Mr. William Halyburton, dated the 27th May, from Warwick; out of which letter I have made this extract:
>
> "'Instead of news I send you some papers of wonders. On Saturday last, it was rumoured in this town, that it rained wheat at Tuchbrooke, a village about two miles from Warwick. Whereupon some of the inhabitants of this town went thither; where they saw great quantities on the way, in the fields, and on the leads of the church, castle, and priory, and upon the hearths of the chimneys in the chambers. And Arthur Mason, coming out of Shropshire, reports, that it hath rained the like in many places of that county. God make us thankful for this miraculous blessing, &.'"

"I brought some papers of these seeds, with this letter, to the Society of Gresham College; who would not enter into any consideration of it, p. 173till they were better informed of the matter of fact. Hereupon, I entreated Mr. Henry Puckering to write to the bailiff of the town of Warwick, to the ministers and physicians, to send us an account of the matter of fact, and their opinions of it. In the bailiff's letter, dated the 3rd of June, I find this report verified; affirming that himself, with the inhabitants of the town, were in a great astonishment at this wonder. But, before the next day of our meeting, I sent for some ivy-berries, and brought them to Gresham College with some of these seeds resembling wheat; and taking off the outward pulp of the ivy-berries, we found in each of the berries four seeds; which were generally concluded by the Society to be the same with those that were supposed and believed by the common people to have been wheat that had been rained; and, that they were brought to those places, where they were found, by starlings; who, of all the birds that we know, do assemble in the greatest numbers; and do, at this time of the year, feed upon these berries; and digesting the outward pulp, they render these seeds by casting, as hawks do feathers and bones."

The remarkable showers already noticed, have p. 174excited much interest and inquiry among learned men, and many superstitious fears among the ignorant; but, there is another description of shower which affords a singular instance of popular observation, being greatly in advance of scientific knowledge. We allude to the showers of stones, called "akrolites," (from two Greek words, signifying the *atmosphere*, and a *stone*); they are also called *Meteorolites*, or *Meteoric stones*.

Writers in all ages have mentioned instances of stony bodies having been seen to fall from the sky. The Chinese and Japanese carefully note down the most striking and remarkable phenomena of nature, believing them to have some connexion with public affairs; and the chronicles of these people are said to contain many notices of the fall of stony bodies from the sky. Until within the last fifty years, however, these accounts have been treated in Europe as idle superstitions; scientific men denying even the probability of such an occurrence. The first scientific man who was bold enough to support the popular opinion, that stones actually do fall from the sky, was Chladni, a German philosopher, who published a pamphlet on the subject in 1794. This did not excite p. 175much attention, until, two years afterwards, a stone weighing fifty-six pounds was exhibited in London, which was said to have fallen in Yorkshire in the December of the preceding year; but, although the fact was attested by several respectable persons, the possibility of such an occurrence was still doubted. It was remarked, however, by Sir Joseph Banks, that this stone was very similar in appearance to one which had been sent to him from Italy, with an account of its having fallen from the clouds. In the year 1799, a number of stones were received by the Royal Society, from Benares, in the East Indies, which were also said to have fallen from the atmosphere, with a minute account of the circumstances attending the fall, which will be presently noticed; and, as these stones appeared to be precisely similar to the Yorkshire stone already noticed, attention was fairly drawn to the subject. In 1802, Mr. Howard published an analysis of a variety of these stones collected from different places; and his researches led to the important conclusion, that they are all composed of the same substances, and in nearly the same proportions. In 1803, a notice was received at Paris, of a shower of stones at L'Aigle in Normandy; p. 176and the Institute of France deputed M. Biot, a well-known and excellent natural philosopher, to examine, on the spot, all the circumstances attending this remarkable event. His account will be noticed presently; but it may here be stated, that the stones he collected, on being analysed, gave results similar to those obtained by Mr. Howard.

The circumstances attending the fall of stones at Krakhut, a village about fourteen miles from the city of Benares, are briefly as follow:—On the 19th December, 1798, a very luminous meteor was observed in the heavens, about eight o'clock in the evening, in the form of a large ball of fire; it was accompanied by a loud noise, resembling that of thunder, which was immediately followed by the sound of the fall of heavy bodies. On examining the ground, it was observed to have been newly torn up in many places; and in these were found stones of a peculiar appearance, most of which had buried themselves to the depth of six inches. At the time the meteor appeared, the sky was perfectly serene, not the smallest vestige of a cloud had been seen since the 11th of the month; nor were any observed for many days p. 177after. It was seen in the western part of the hemisphere, and was visible only a short time. The light from it was so great, as to cast a strong shadow from the bars of a window upon a dark carpet. Mr. Davis, the judge and magistrate of the district, affirmed, that in brilliancy it equalled the brightest moonlight. Both he and Mr. Erskine were induced to send persons in whom they could confide to the spot where this shower of stones is reported to have taken place, and thus obtained additional evidence of the phenomena, together with several of the stones which had penetrated about six inches into fields recently watered. Mr. Maclane, a gentleman who resided near Krakhut, presented Mr. Howard with a portion of a stone which had been brought to him the morning after its fall by the person who was on duty at his house, and through the roof of whose hut it had passed, and buried itself several inches in the floor, which was of consolidated earth. Before it was broken it must have weighed upwards of two pounds.

M. Biot's summary of the evidence collected by him respecting the great shower of stones which fell at Aigle, in Normandy, is as follows:—

> p. 178"On Tuesday, 26th April, 1803, about one o'clock, p.m., the weather being serene, there was observed from Caen, Pont d'Audemer, and the environs of Alengon, Falaise, and Verneuil, a fiery globe, of a very brilliant splendour, and which moved in the atmosphere with great rapidity. Some moments after, there was heard at Aigle, and in the environs of that town, in the extent of more than thirty leagues in every direction, a violent explosion, which lasted five or six minutes. At first there were three or four reports like those of a cannon, followed by a kind of discharge which resem-

bled the firing of musketry; after which, there was heard a dreadful rumbling, like the beating of a drum. The air was calm and the sky serene, except a few clouds, such as are frequently observed. This noise proceeded from a small cloud which had a rectangular form; the largest side being in a direction from east to west. It appeared motionless all the time that the phenomenon lasted; but the vapours of which it was composed, were projected momentarily from different sides, by the effect of successive explosions. This cloud was about half a league to the north-north-west of the town of Aigle. p. 179It was at a great elevation in the atmosphere; for, the inhabitants of two hamlets, a league distant from each other, saw it at the same time above their heads. In the whole canton over which this cloud was suspended, there was a hissing noise, like that of a stone discharged from a sling; and a great many mineral masses, exactly similar to those distinguished by the name of 'meteor-stones,' were seen to fall. The district in which these masses were projected, forms an elliptical extent of about two leagues and a half in length, and nearly one in breadth, the greatest dimension being in a direction from south-east to north-west; forming a declination of about 22 degrees. This direction, which the meteor must have followed, is exactly that of the magnetic meridian, which is a remarkable result. The greatest of these stones fell at the south-eastern extremity of the large axis of the ellipse, the middle-sized in the centre, and the smaller at the other extremity. Hence it appears, that the largest fell first, as might naturally be supposed. The largest of all those that fell, weighs seventeen pounds and a half. The smallest which I have seen, weighs about two *gros*, (a thousandth part of the last.) The number p. 180of all those which fell, is certainly above two or three thousand."

Meteoric stones have been known to commit great injury in their fall. In July, 1790, a very bright fire-ball, luminous as the sun, of the size of an ordinary balloon, appeared near Bourdeaux, which, after filling the inhabitants with alarm, burst, and disappeared. A few days after, some peasants brought stones into the town, which they said had fallen from the meteor; but, the philosophers to whom they offered them laughed at their statements. One of these stones, fifteen inches in diameter, broke through the roof of a cottage, and killed a herdsman and a bullock. In 1810, a great stone fell at Shahabad, in India. It burnt a village, and killed several people.

The fall of meteoric stones is more frequent than would be supposed. Chaldni has compiled a Catalogue of all recorded instances from the earliest times. Of these, twenty-seven are previous to the Christian era; thirty-five from the beginning of the first to the end of the fourteenth century; eighty-nine from the beginning of the fifteenth to the beginning of the present century; from which time, since the attention of scientific men has p. 181been directed to the subject, above sixty cases have been recorded. These are, doubtless, but a small proportion of the whole amount of meteoric showers which have fallen, when the small extent of surface occupied by those capable of recording the event is compared with the wide expanse of the ocean, the vast uninhabited deserts, mountains, and forests, and the countries occupied by savage nations.

Meteoric stones have generally a broken, irregular surface, coated with a thin black crust, like varnish. When broken, they appear to have been made up of a number of small spherical bodies of a grey colour, imbedded in a gritty substance, and often interspersed with yellow spots. A considerable proportion of iron is found in all of them, partly in a malleable state, partly in that of an oxide, and always in combination with a rather scarce metal called nickel; [181] the earths silica, and p. 182magnesia, and sulphur, form the other chief ingredients; but, the earths alumina and lime, the metals manganese, chrome, and cobalt, together with carbon, soda, and water, have also been found in small quantities, but not in the same specimens. No substance with which chemists were previously unacquainted, has ever been found in them; but no combination, similar to that in meteoric stones, has ever been met with in geological formations, or among the products of any volcano. They are sometimes very friable, sometimes very hard; and some that are friable when they first fall, become hard afterwards. When taken up soon after their fall they are extremely hot. They vary in weight from two drams to several hundred pounds. Meteoric stones have fallen in all climates, in every part of the earth, at all seasons, in the night and in the day.

The meteoric stones already noticed, are not the only metallic bodies which are supposed to fall from the sky. In many parts of the earth masses of malleable iron, often of vast size, have been found. An immense mass seen by Pallas, in Siberia, was discovered at a great height on a mountain of slate, near the river Jenesei. The Tartars held p. 183it in great veneration, as having fallen from heaven. It was removed in the year 1749, to the town of Krasnojarsk, by the inspector of iron mines. The mass, which weighed about 1,400 pounds, was irregular in form, and cellular, like a sponge. The iron was tough and malleable, and was found to contain nickel, silica, magnesia, sulphur, and chrome. Another enormous mass of meteoric iron was found in South America, about the year 1788. It lay in a vast plain, half sunk in the ground, and was supposed, from its size and the known weight of iron, to contain upwards of thirteen tons. Specimens of this mass are now in the British Museum, and have been found to contain 90 per cent. of iron and 10 of nickel. Many other masses of iron might be mentioned, which, from the places in which they are found, and from their composition, leave no doubt as to their being of meteoric origin. The only instance, on record, of iron having been actually seen to fall from the atmosphere, is that which took place at Agram in Croatia, on the 26th May, 1751. About six o'clock in the evening, the sky being quite clear, a ball of fire was seen, which shot along, with a hollow noise, from west to east, and, after a loud explosion accompanied by a great p. 184smoke, two masses of iron fell from it in the form of chains welded together.

It is, perhaps, impossible, in the present state of our knowledge, to account for the origin of these remarkable bodies. Some have supposed them to have been shot out from volcanoes belonging to our earth; but this theory is opposed by the fact that no substance, resembling akrolites, has ever been found in or near any volcano; and they fall from a height to which no volcano can be supposed to have projected them, and still less to have given them the horizontal direction in which they usually move. Another supposition is, that these masses are formed in the atmosphere; but it is almost ridiculous to imagine a body, weighing many tons, to be produced by any chemical or electrical forces in the upper regions of the air. A third explanation is, that they are bodies thrown out by the volcanoes, which are known to exist in the moon, with such force as to bring them within the sphere of the earth's attraction. This notion was supported by the celebrated astronomer and mathematician La Place. He calculated that a body projected from the moon with the velocity of 7771 feet in the first second, would reach our earth in p. 185about two days and a half. But other astronomers are of opinion, that the known velocity of some meteors is too great to admit of the possibility of their having come from the moon. The theory which agrees best with known facts and the laws of nature, is that proposed by Chladni, namely, that the meteors are bodies moving in space, either masses of matter as originally created, or fragments separated from a larger mass of a similar nature. This view has also been supported by Sir Humphrey Davy, who says, "The luminous appearances of shooting-stars and meteors cannot be owing to any inflammation of elastic fluids, but must depend upon the ignition of solid bodies. Dr. Halley calculated the height of a meteor at ninety miles; and the great American meteor, which threw down showers of stones, was estimated at seventeen miles high. The velocity of motion of these bodies must, in all cases, be immensely great, and the heat produced by the compression of the most rarefied air from the velocity of motion, must be, probably, sufficient to ignite the mass; and all the phenomena may be explained, if *falling stars* be supposed to be small bodies moving round the earth in very eccentric orbits, which become ignited only when p. 186they pass with immense velocity through the upper region of the atmosphere; and if the meteoric bodies which throw down stones with explosions, be supposed to be similar bodies which contain either combustible or elastic matter."

This chapter ought not to be concluded without a short notice of that remarkable rain known to geologists as "fossil rain." In the new red-sandstone of the Storeton quarries, impressions of the footprints of ancient animals have been discovered; and in examining some of the slabs of stone extracted at the depth of above thirty feet,

Mr. Cunningham observed "that their under surface was thickly covered with minute hemispherical projections, or casts in relief of circular pits, in the immediately subjacent layers of clay. The origin of these marks, he is of opinion, must be ascribed to showers of rain which fell upon an argillaceous beach exposed by the retiring tide, and their preservation to the filling up of the indentations by sand. On the same slabs are impressions of the feet of small reptiles, which appear to have passed over the clay previously to the shower, since the foot-marks are also indented with circular pits, but to a less degree; and the difference Mr. Cunningham p. 187explains by the pressure of the animal having rendered these portions less easily acted upon." The preservation of these marks has been explained by supposing dry sand, drifted by the wind, to have swept over and filled up the footprints, rain-pits, and hollows of every kind, which the soft argillaceous surface had received.

The frontispiece to the present chapter (p. 156), represents a slab of sandstone containing impressions of the foot of a bird and of rain drops. This slab is from a sandstone basin near Turner's Falls, a fine cataract of the Connecticut river in the State of Massachusetts, and is described by Dr. Deane in a recent number of the American Journal of Science. "It is rare," says that gentleman, to "find a stratum containing these footprints exactly as they were made by the animal, without having suffered change. They are usually more or less disturbed or obliterated by the too soft nature of the mud, the coarseness of the materials, and by many other circumstances which we may easily see would deface them, so that although the general form of the foot may be apparent, the minute traces of its appendages are almost invariably lost. In general, except in thick-toed species, we p. 188cannot discover the distinct evidences of the structure of the toes, each toe appearing to be formed of a single joint, and seldom terminated by a claw. But, a few specimens hitherto discovered at this locality completely developed the true characters of the foot, its ranks of joints, its claws and integuments. So far as I have seen, the faultless impressions are upon shales of the finest texture with a smooth glossy surface, such as would retain the beautiful impressions of rain drops. This kind of surface containing foot-marks is exceedingly rare: I have seen but few detached examples; recently it has been my good fortune to recover a stratum, containing in all more than one hundred most beautiful impressions of the feet of four or five varieties of birds, the entire surface being also pitted by a shower of fossil rain-drops. The slabs are perfectly smooth on the inferior surface, and are about two inches in thickness.

"The impression of a medallion is not more sharp and clear than are most of these imprints, and it may be proper to observe, that this remarkable preservation may be ascribed to the circumstance, that the entire surface of the stratum was incrusted with a layer of micaceous sandstone, adhering so firmly that it would not cleave off, thereby requiring the laborious and skilful application of the chisel. The appearance of this shining layer which is of a gray colour, while the fossil slab is a dark red, seems to carry the probability that it was washed or blown over the latter while in a state of loose sand, thus filling up the foot-prints and rain-drops, and preserving them unchanged until the present day — unchanged in the smallest particular, so far as relates merely to configuration, nothing being obliterated; the precise form of the nails, or claws, and joints, and in the deep impressions of the heel bone, being exquisitely preserved."

The small slab figured at p. 156 is described as being an incomparable specimen. "For purity of impression it is unsurpassed, and the living reality of the rain-drops, the beautiful colour of the stone, its sound texture and lightness, renders it a fit member for any collection of organic remains."

p. 191CHAPTER VIII.

common sayings respecting the weather—saint swithin's-day—signs of rain or of fair weather derived from the appearance of the sun—from that of the moon—from the stars—from the sky—from the distinctness of sounds—from the rising of smoke—from the peculiar actions of plants and animals—prognostics noticed by sir humphrey davy—signs of rain collected by dr. jenner—north american rain-makers—incident related by catlin—rain-doctors of southern africa—rain-doctors of ceylon—superstitions giving way to the teaching of missionaries—conclusion.

There are many proverbial sayings among country people concerning the state of the weather, which, having been derived from long observation, have become axioms, and were designated by Bacon "the philosophy of the people." These prognostics are being set aside by the more certain lights of science, but there is no doubt that many natural objects may indicate symptoms of change in the atmosphere before any actually takes place in it to such an extent as to affect our senses. Some of p. 192these prognostics are of a general character applying to all seasons, and there are others which apply only to a particular season; but they may all be derived from appearances of the heavenly bodies and of the sky, the state of meteorological instruments, and the notions and habits of certain plants and animals. The author of the "Journal of a Naturalist" has some good observations on this subject. He says:—

> "Old simplicities, tokens of winds and weather, and the plain observances of human life, are everywhere waning fast to decay. Some of them may have been fond conceits; but they accorded with the ordinary manners of the common people, and marked times, seasons, and things, with sufficient truth for those who had faith in them. Little as we retain of these obsolete fancies, we have not quite abandoned them all; and there are yet found among our peasants a few, who mark the blooming of the large water-lily (*lilium candidum*), and think that the number of its blossoms on a stem

will indicate the price of wheat by the bushel for the ensuing year, each blossom equivalent to a shilling. We expect a sunny day too, when the pimpernel (*anagallis arvensis*) fully expands its blossoms; a p. 193dubious, or a moist one, when they are closed. In this belief, however, we have the sanction of some antiquity to support us. Sir F. Bacon records it; Gerarde notes it as a common opinion entertained by country people above two centuries ago; and I must not withhold my own faith in its veracity, but say that I believe this pretty little flower to afford more certain indication of dryness or moisture in the air than any of our hygrometers do. But if these be fallible criterions, we will notice another that seldom deceives us. The approach of a sleety snow-storm, following a deceitful gleam in spring, is always announced to us by the loud untuneful voice of the missel-thrush (*turdus viscivorus*) as it takes its stand on some tall tree, like an enchanter calling up the gale. It seems to have no song, no voice, but this harsh predictive note; and it in great measure ceases with the storms of spring. We hear it occasionally in autumn, but its voice is not then prognostic of any change of weather. The missel-thrush is a wild and wary bird, keeping generally in open fields and commons, heaths and unfrequented places, feeding upon worms and insects. In severe weather it approaches our plantations and shrubberies, to feed p. 194on the berry of the mistletoe, the ivy, or the scarlet fruit of the holly or the yew; and, should the redwing or the fieldfare presume to partake of these with it, we are sure to hear its voice in clattering and contention with the intruders, until it drives them from the place, though it watches and attends, notwithstanding, to its own safety."

But before we notice more in detail the natural prognostics of the weather, it is desirable to speak of a superstition which is widely spread among all classes, in the town as well as in the country. The superstition referred to, is that connected with St. Swithin's-day, and is well expressed in a Scotch proverb: —

"Saint Swithin's-day, gif ye do rain,
For forty days it will remain;
Saint Swithin's-day, an ye be fair,
For forty days 'twill rain nae mair."

This superstition originated with Swithin, or Swithum, bishop of Winchester, who died in the year 868. He desired that he might be buried in the open churchyard, "where the drops of rain might wet his grave;" "thinking," says Bishop Hall, "that no vault was so good to cover his grave as that of heaven." But when Swithin was canonized p. 195the monks resolved to remove his body into the choir of the church. According to tradition, this was to have been done on the 15th of July; but it rained so violently for forty days that the design was abandoned. Mr. Howard remarks, that the tradition is so far valuable, as it proves that the summers in the southern part of our island were subject, a thousand years ago, to occasional heavy rains, in the same way as at present. This accurate observer has endeavoured to ascertain how far the popular notion is borne out by the fact. In 1807 and 1808, it rained on St. Swithin's-day, and a dry season followed. In 1818 and 1819, it was dry on the 15th, and a very dry season followed. The other summers, occurring between 1807 and 1819, seem to show, "that in a majority of our summers, a showery period which, with some latitude as to time and local circumstances, may be admitted to constitute daily rain for forty days, does come on about the time indicated by the tradition of St. Swithin."

But in these calculations, it is necessary to bear in mind that the change of style has very much interfered with St. Swithin. With the day allowed in the closing year of the last century, St. Swithin's p. 196day is how thirteen days earlier in the calendar than it would have been by the old style. Thus the true St. Swithin's-day, according to the tradition, is about the 28th of July, and not the 15th, as set down in the present calendar. There must, therefore, be a considerable difference as to the rains and this day.

We now proceed to collect a number of prognostics connected with the appearances of the heavenly bodies and of the sky; they are the result of long experience, but at the same time it is necessary to caution our readers against attaching much importance to them.

When the sun rises red, wind and rain may be expected during the day; but when he rises unclouded, attended by a scorching heat, cloudiness and perhaps rain will ensue before mid-day. When he rises clouded, with a few grey clouds, they will soon dissipate, and a fine day will follow. When his light is dim, vapour exists in the upper regions of the air, and may be expected to descend shortly after in the form of dense clouds. When his light, after rain, is of a transparent watery hue, rain will soon fall again. When his direct rays have a scorching and weakening effect on the body p. 197throughout the greater part of the day, the next day will be cloudy, and perhaps rainy. When the sun is more or less obscured by a thicker or thinner cirro-stratus cloud, and when he is said to be *wading* in the cloud, rain may come—if the cloud indicates rain it will come. A halo surrounding the disc of the sun is almost always sure to precede rain. A red sunset without clouds indicates a doubt of fair weather; but a fine day may be expected after a red sunset in clouds. A watery sunset, diverging rays of light, either direct from the sun or from behind a cloud, is indicative of rain. After a dull black sunset rain may be expected.

It is a common saying among country people,—

"An evening red, or a morning grey,
Doth betoken a bonnie day;
In an evening grey and a morning red,
Put on your hat, or yell weet your head."

There are not many prognostics connected with the appearances of the moon. The changes of the moon produce greater effects than at any other period. With a clear silvery aspect fair weather may be expected. A pale moon always indicates rain, and a red one wind. Seeing the "old moon p. 198in the new one's arms," is a sign of stormy weather. Seeing the new moon very young, "like the paring of a nail," also indicates wet; but when the horns of the new moon are blunt, they indicate rain, and fair weather when sharp. It is truly said:

"In the wane of the moon,
A cloudy morning bodes a fair afternoon."

And also

'New moon's mist
Never dies of thirst.'

Halos and coronf are oftener seen about the moon than the sun, and they indicate rain.

The stars appearing dim indicate rain. Very few stars seen at one time, when there is no frost, indicate a similar result.

When the sky is of deeply-coloured blue, it indicates rain. If distant objects appear very distinct and near through the air, it indicates rain. When the air feels oppressive to walk in, rain will follow; when it feels light and pleasant, fair weather will continue.

When distant sounds are distinctly heard through the air in a calm day, such as the tolling of bells, barking of dogs, talking of people, waterfalls, or rapids over mill-dams, the air is loaded with vapour, p. 199and rain may be expected. The sea is often heard to roar, and loudest at night, as also the noise of a city, when a cloud is seen suspended a very short way above head.

If smoke rise perpendicularly upwards from chimneys in calm weather, fair weather may be expected to continue; but if it fall toward and roll along the ground, not being easily dispersed, rain will ensue.

Many of the above prognostics, as well as some of those relating to animals, are thus noticed by Sir Humphrey Davy, in his "Salmonia, or Days of Fly-fishing." The conversation is between Halieus, a fly-fisher; Poietes, a poet; Physicus, a man of science; and Ornither, a sportsman.

> "*Poiet.* I hope we shall have another good day to-morrow, for the clouds are red in the west.
>
> *Phys.* I have no doubt of it; for the red has a tint of purple.

Hal. Do you know why this tint portends fine weather?

Phys. The air, when dry, I believe, refracts more red or heating rays; and as dry air is not perfectly transparent, they are again reflected in the horizon. I have generally observed a coppery p. 200or yellow sun-set to foretell rain; but, as an indication of wet weather approaching, nothing is more certain than a halo round the moon, which is produced by the precipitated water; and the larger the circle, the nearer the clouds, and consequently the more ready to fall.

Hal. I have often observed that the old proverb is correct—

'A rainbow in the morning is the shepherd's warning;
A rainbow at night is the shepherd's delight'

Can you explain this omen?

Phys. A rainbow can only occur when the clouds containing or depositing the rain are opposite the sun,—and in the evening the rainbow is in the east, and in the morning in the west. As, therefore, our heavy rains in this climate are usually brought by the westerly wind, a rainbow in the west indicates that the bad weather is on the road, by the wind, to us; whereas, the rainbow in the east proves that the rain in these clouds is passing from us.

Poiet. I have often observed that when the swallows fly high, fine weather is to be expected or continued; but when they fly low, and close to p. 201the ground, rain is almost surely approaching. Can you account for this?

Hal. Swallows follow the flies and gnats, and flies and gnats usually delight in warm strata of air; and as warm air is lighter, and usually moister than cold air, when the warm strata of air are high, there is less chance of moisture being thrown down from them by the mixture with cold air; but when the warm and moist air is close to the surface, it is almost certain that, as the cold air flows down into it, a deposition of water will take place.

Poiet. I have often seen sea-gulls assemble on the land, and have almost always observed that very stormy and rainy weather was approaching. I conclude that these animals, sensible of a current of air approaching from the ocean, retire to the land to shelter themselves from the storm.

Orn. No such thing. The storm is their element, and the little petrel enjoys the heaviest gale; because, living on the smaller sea-insects, he is sure to find his food in the spray of a heavy wave; and you may see him flitting above the edge of the highest surge. I believe that the reason of this migration of sea-gulls, and other sea-birds, to the land, is their security of finding food; and they p. 202may be observed, at this time, feeding greedily on the earth-worms and larvf driven out of the ground by severe floods; and the fish on which they prey in fine weather in the sea, leave the surface, and go deeper in storms. The search after food, as we have agreed on a former occasion, is the principal cause why animals change their places. The different tribes of the wading birds always migrate when rain is about to take place; and I remember once, in Italy, having been long waiting, in the end of March, for the arrival of the double snipe in the Campagna of Rome, a great flight appeared on the 3rd of April, and the day after heavy rain set in, which greatly interfered with my sport. The vulture, upon the same principle, follows armies; and I have no doubt that the augury of the ancients was a good deal founded upon the observation of the instincts of birds. There are many superstitions of the vulgar owing to the same source.

For anglers, in spring, it is always unlucky to see single magpies,—but *two* may always be regarded as a favourable omen; and the reason is, that in cold and stormy weather one magpie alone leaves the nest in search of food, the other remaining sitting upon the eggs or the young ones; but when p. 203two go out together, it is only when the weather is warm and mild, and favourable for fishing.

Poiet. The singular connexions of causes and effects to which you have just referred, makes superstition less to be wondered at, particularly amongst the vulgar; and when two facts, naturally unconnected, have been accidentally coincident, it is not singular that this coincidence should have been observed and registered, and that omens of the most absurd kind should be trusted in. In the west of England, half a century ago, a particular hollow noise on the sea-coast was referred to a spirit or goblin, called Bucca, and was supposed to foretell a shipwreck. The philosopher knows that sound travels much faster than currents in the air; and the sound always foretold the approach of a very heavy storm, which seldom takes place on that wild and rocky coast without a shipwreck on some part of its extensive shores, surrounded by the Atlantic."

Dr. Jenner has collected in the following amusing lines a large number of the natural prognostics of rain. They are said to have been addressed to a lady, who asked the Doctor if he thought it would rain to-morrow.

p. 204"The hollow winds begin to blow,
The clouds look black, the glass is low;
The soot falls down, the spaniels sleep,
And spiders from their cobwebs peep:
Last night the sun went pale to bed,
The moon in halos hid her head:
The boding shepherd heaves a sigh,

For, see! a rainbow spans the sky:
The walls are damp, the ditches smell,
Closed is the pink-eyed pimpernel;
Hark! how the chairs and tables crack;
Old Betty's joints are on the rack;
Loud quack the ducks, the peacocks cry,
The distant hills are seeming nigh.
How restless are the snorting swine,—
The busy flies disturb the kine.
Low o'er the grass the swallow wings;
The cricket, too, how loud it sings:
Puss on the hearth with velvet paws,
Sits smoothing o'er her whisker'd jaws.
Through the clear stream the fishes rise,
And nimbly catch the incautious flies:
The sheep were seen at early light
Cropping the meads with eager bite.
Though June, the air is cold and chill;
The mellow blackbird's voice is still.
The glow-worms, numerous and bright,
Illum'd the dewy dell last night.
At dusk the squalid toad was seen,
Hopping, and crawling o'er the green.
The frog has lost his yellow vest,
And in a dingy suit is dressed.
p. 205The leech, disturb'd, is newly risen,
Quite to the summit of his prison.
The whirling winds the dust obeys,
And in the rapid eddy plays;
My dog, so alter'd in his taste,
Quits mutton-bones on grass to feast;
And see yon rooks, how odd their flight!
They imitate the gliding kite,
Or seem precipitate to fall,
As if they felt the piercing ball:—
'Twill surely rain,—I see with sorrow,
Our jaunt must be put off to-morrow."

Uncivilized nations often entertain the absurd notion that certain individuals can command the rain whenever they please. Much honour is shown to persons supposed to possess this power, for they are considered as having some mysterious intercourse with heaven. Catlin gives a striking instance of this superstition as it exists among the Mandans of North America. These people raise a great deal of corn; but their harvests are sometimes destroyed by long-continued drought. When threatened with this calamity, the women (who have the care of the patches of corn) implore their lords to intercede for rain; and accordingly the chiefs and doctors assemble to deliberate on the case. When they have decided that it is necessary p. 206to produce rain, they wisely delay the matter for as many days as possible; and it is not until further urged by the complaints and entreaties of the women, that they begin to take the usual steps for accomplishing their purpose. At length they assemble in the council-house with all their apparatus about them—with abundance of wild sage and aromatic herbs, to burn before the "Great Spirit." On these occasions the lodge is closed to all except the doctors and some ten or fifteen young men, the latter being the persons to whom the honour of making it rain, or the disgrace of having failed in the attempt, is to belong.

After having witnessed the conjurations of the doctors inside the lodge, these young men are called up by lot, one at a time, to spend a day on the top of the lodge, and to see how far their efforts will avail in producing rain; at the same time the smoke of the burning herbs ascends through a hole in the roof. On one of these occasions, when all the charms were in operation, and when three young men had spent each his day on the lodge in ineffectual efforts to bring rain, and the fourth was engaged alternately addressing the crowd of villagers and the spirits of the air, but p. 207in vain, it so happened that the steam-boat "Yellow Stone," made her first trip up the Missouri river, and about noon approached the village of the Mandans. Catlin was a passenger on this boat; and helped to fire a salute of twenty guns of twelve pounds calibre, when they first came in sight of the village, which was at some three or four miles distance. These guns introduced a new sound into the country, which the Mandans naturally enough supposed to be thunder. "The young man upon the lodge, who turned it to good account, was gathering fame in rounds of applause, which were repeated and echoed through the whole village; all eyes were centred upon him—chiefs envied him—mothers' hearts were beating high, whilst they were decorating and leading up their fair daughters to offer him in marriage on his signal success. The medicine-men had left the lodge, and came out to bestow upon him the envied title of 'medicine-man,' or 'doctor,' which he had so deservedly won—wreaths were prepared to decorate his brows, and eagle's plumes and calumets were in readiness for him—his enemies wore on their faces a silent gloom and hatred; and his old sweethearts who had cast him off, gazed p. 208intensely upon him, as they glowed with the burning fever of repentance. During all this excitement, Wak-a-dah-ha-hee (or the white buffalo's hair) kept his position, assuming the most commanding and threatening attitudes; brandishing his shield in the direction of the thunder, although there was not a cloud to be seen, until he (poor fellow) being elevated above the rest of the village, espied, to his inexpressible amazement, the steamboat ploughing its way up the windings of the river below, puffing her steam from her pipes, and sending forth the thunder from a twelve-pounder on her deck. 'The white Buffalo's hair' stood motionless, and turned pale; he looked awhile, and turned to the chief and to the multitude, and addressed them with a trembling lip—'My friends, we will get no rain!—there are, you see, no clouds; but my medicine is great—I have brought a *thunder-boat*! look and see it! the thunder you hear is out of her mouth, and the lightning which you see is on the waters!'

"At this intelligence, the whole village flew to the tops of their wigwams, or to the bank of the river, from whence the steamer was in full view, and ploughing along to their utter dismay and p. 209confusion. In this promiscuous throng, chiefs, doctors, women, children, and dogs, were mingled, Wak-a-dah-ha-hee having descended from his high place to mingle with the frightened throng. Dismayed at the approach of so strange and unaccountable an object, the Mandans stood their ground but a few moments; when, by an order of the chiefs, all hands were ensconced within the piquets of their village, and all the warriors armed for desperate self-defence. A few moments brought the boat in front of the village, and all was still and quiet as death; not a Mandan was to be seen upon the banks. The steamer was moored, and three or four of the chiefs soon after walked boldly down the bank, and on to her deck, with a spear in one hand, and a calumet, or pipe of peace in the other. The moment they stepped on board, they met (to their great surprise and joy) their old friend Major Sanford, their agent, which circumstance put an instant end to all their fears."

It was long, however, before the rain-maker could be persuaded to come forward, or to listen to the assurance that his medicine had nothing whatever to do with the arrival of the ship. Unwilling to lose the fame of having produced p. 210such a phenomenon, he continued to assert that he knew of its coming, and by his magic had caused it to approach. But he was little regarded in the universal bustle and gossip which was going on respecting the mysteries of the "thunder-boat."

Meanwhile the day passed on, and towards evening a cloud began to rise above the horizon. Wak-a-dah-ha-hee no sooner observed this, than, with shield on his arm and bow in hand, he was again upon the lodge. "Stiffened and braced to the last sinew, he stood with his face and his shield presented to the cloud, and his bow drawn. He drew the eyes of the whole village upon him, as he vaunted forth his superhuman powers; and at the same time commanded the cloud to come nearer, that he might draw down its contents upon their heads and the corn-fields of the Mandans. In this wise he stood, waving his shield over his head, stamping his foot, and frowning as he drew his bow and threatened the heavens, commanding it to rain—his bow was bent, and the arrow drawn to its head, was sent to the cloud, [210] and he exclaimed, 'My friends, it is done! Wak-a-dah-ha-hee's arrow has entered that black cloud, and p. 211the Mandans will be wet with the water of the skies!' His predictions were true—in a few moments the cloud was over the village, and the rain fell in torrents. He stood for some time wielding his weapons, and boasting of the efficacy of his *medicine* to those who had been about him, but were now driven to the shelter of their wigwams; and descended from his high place (in which he had been perfectly drenched) prepared to receive the honours and homage that were due to one so potent in his mysteries; and to receive the style and title of *medicine-man*." Catlin further informs us, that when the Mandans undertake to make it rain, they always succeed, for their ceremonies never stop until rain begins to fall: and also, that he who has once made it rain never attempts it again; his medicine is undoubted—and on future occasions of the kind he stands aloof, giving an opportunity to other young men who are ambitious to signalize themselves in the same way.

A superstition similar to that of the Mandans prevails also among the Caffers of Southern Africa, and among the natives of Ceylon. The Caffer chiefs, attended by their warriors, proceed with much ceremony, and laden with presents, to the dwelling of the rain-doctor, where a grand feast is held while certain charms are in process. The impostor at length dismisses his guests with a variety of instructions, on the due observance of which the success of their application is to depend. These instructions are generally of the most trifling kind: they are to travel home in perfect silence; or they are not to look back; or they are to compel every one they meet to turn back and go home with them. Should rain happen to fall, the credit is given to the rain-doctor; but should the drought continue, the fault is laid upon the failure of the applicants to fulfil these instructions with sufficient exactness.

Major Forbes gives an account of an old rain-doctor in Ceylon, who had plied a lucrative trade for many years, and at length wished to retire from business. But the people were highly incensed at the idea of losing his services, especially as a most distressing drought was at that time the scourge of the land. So persuaded were they of his powers, that they all agreed, that when required to do so by a whole village, he should be compelled to furnish rain in sufficient quantities; and that if he was insensible to rewards, he should be tormented with thorns or beaten into compliance. In vain did the poor old impostor at length declare the truth, and assure the people that he had no power whatever to make it rain. They treated his words with disdain, and dragged their victim from village to village, inflicting stripes at every halt. Even the chief of the district had determined on having rain by force, if fair means should fail, and ordered the rain-doctor to be taken to the village where rain was most required. On his way thither he was so fortunate as to meet with Major Forbes, who took him under his protection, and probably saved his life, though not without some difficulty, for it so happened that a few slight showers fell near his own village, while all the rest of the neighbourhood was suffering the extremity of drought.

Melancholy indeed is the condition of these poor people; in utter ignorance of the source of all the providential mercies bestowed upon them, and, therefore, made the dupes and credulous followers of knaves and impostors of every kind!

In some cases, however, the missionaries have happily succeeded in opening the eyes of the p. 214deluded people to the cheat which is practised on them. One of the most intelligent of the Caffers of Southern Africa, having been led to suspect the integrity of the rain-maker, visited Mr. Shaw, and told him of his determination to have the question set at rest, whether or no the rain-maker could produce rain. He had summoned the rain-maker to meet Mr. Shaw in an open plain, when all the Caffers of the surrounding kraals were to be present to decide the affair. Accordingly, at the appointed time and place, thousands of Caffers from the neighbouring country assembled in their war-dresses. Mr. Shaw, being confronted with a celebrated rain-maker, declared publicly that God alone gave rain; and then offered to present the rain-maker with a team of oxen if he should succeed in making it rain within a certain specified time. This was agreed to; the rain-maker began his ceremonies, which are said to have been well calculated to impose upon an ignorant and superstitious people. The time having expired without any signs of rain, the chief who had called together the meeting asked the rain-maker why he had so long imposed upon them? The rain-maker complained that he had not been paid p. 215well enough for his rain; and appealed to all present, whether rain had not always been produced when he had been properly paid. Mr. Shaw then pointed out some half-famished cattle belonging to the rain-maker, which were seen on a neighbouring hill starving for want of pasturage, and remarked, that if he really possessed his boasted skill, he would not have neglected his own interests. To this the rain-maker cleverly replied, "I never found a difficulty in making rain until *he* (pointing to Mr. Shaw) came among us; but now, no sooner do I collect the clouds, and the rain is about to fall, than immediately there begins a sound of *ting, ting, ting,* (alluding to the chapel-bell,) which puts the clouds to flight, and prevents the rain from descending on your land." Mr. Shaw was not able to tell what effect this ingenious excuse had upon the majority of the Caffers, but he had the satisfaction of knowing that the intelligent chief, who consulted him on the subject, never *bought* any more rain.

p. 216Already Published in this Series.

I. – THE SNOW STORM.
II. – THE FROZEN STREAM.
III. – THE RAIN CLOUD.

Shortly will be Published.

IV.—THE DEW DROP.
V.—THE THUNDER STORM.
VI.—THE TEMPEST.

Footnotes:

[18] Physico-Theology by the Rev. Wm. Derham.

[55] See Frontispiece to this Chapter, p. 36 .

[85] See Frontispiece to this Chapter, p. 74 .

[133] This plan was brought before the notice of the British Association for the advancement of Science in the year 1840.

[155] Harvey's Meteorology, in the Encyclopfdia Metropolitana.

[181] One of the stones which fell at L'Aigle, on being analysed by Thenard, gave —

Silica	46 per cent.
Magnesia	10
Iron	45
Nickel	2
Sulphur	5

[210] See Frontispiece to this Chapter, p. 190.

www.ingramcontent.com/pod-product-compliance
Lightning Source LLC
Chambersburg PA
CBHW031424210526
45464CB00005B/2040